E-MRS 2013 Spring Meeting, Symposium G: Alternative Approaches of SiC and Related Wide Bandgap Materials in Light Emitting and Solar Cell Applications

IOP Conference Series: Materials Science and Engineering Volume 56

Strasbourg, France
27 – 31 May 2013

ISBN: 978-1-63439-867-1
ISSN: 1757-8981

Printed from e-media with permission by:

Curran Associates, Inc.
57 Morehouse Lane
Red Hook, NY 12571

Some format issues inherent in the e-media version may also appear in this print version.

Printed by Curran Associates, Inc. (2015)

For permission requests, please contact the Institute of Physics
at the address below.

Institute of Physics
Dirac House, Temple Back
Bristol BS1 6BE UK

Phone: 44 1 17 929 7481
Fax: 44 1 17 920 0979

techtracking@iop.org

Additional copies of this publication are available from:

Curran Associates, Inc.
57 Morehouse Lane
Red Hook, NY 12571 USA
Phone: 845-758-0400
Fax: 845-758-2634
Email: curran@proceedings.com
Web: www.proceedings.com

TABLE OF CONTENTS

IOP Conf. Series: Materials Science and Engineering **56** (2014) 011001 doi:10.1088/1757-899X/56/1/011001

Preface

Alternative approaches of SiC & related wide bandgap materials in light emitting & solar cell applications

Peter Wellmann[1], Mikael Syväjärvi[2] and Haiyan Ou[3]

[1]University of Erlangen-Nuremberg, Materials Department 6, Martensstr. 7, 91058 Erlangen, Germany, Phone number +49 09131 8527635,
Fax number +49 09131 8528495
[2]Linköping University, Department Physics, Chemistry and Biology, SE-58183 Linköping, Sweden, Phone number: +46 13 285708, Fax number: +46 13 142337
[3]Technical University of Denmark, Department of photonics engineering, Building 343, Oersteds Plads, DK-2800, Kgs. Lyngby, Denmark,
Phone number: +45 4525 6374, Fax number: +45 4593 6581,

E-mail address: peter.wellmann@ww.uni-erlangen.de
E-mail address: mikael.syvajarvi@ifm.liu.se
E-mail address: haou@fotonik.dtu.dk

Materials for optoelectronics give a fascinating variety of issues to consider. Increasingly important are white light emitting diode (LED) and solar cell materials. Profound energy savings can be done by addressing new materials.

White light emitting diodes are becoming common in our lighting scene. There is a great energy saving in the transition from the light bulb to white light emitting diodes via a transition of fluorescent light tubes. However, the white LEDs still suffer from a variety of challenges in order to be in our daily use. Therefore there is a great interest in alternative lighting solutions that could be part of our daily life. All materials create challenges in fabrication. Defects reduce the efficiency of optical transitions involved in the light emitting diode materials. The donor-acceptor co-doped SiC is a potential light converter for a novel monolithic all-semiconductor white LED. In spite of considerable research, the internal quantum efficiency is far less than theoretically predicted and is likely a fascinating scientific field for studying materials growth, defects and optical transitions.

Still, efficient Si-based light source represents an ongoing research field in photonics that requires high efficiency at room temperature, wavelength tuning in a wide wavelength range, and easy integration in silicon photonic devices. In some of these devices, rare earth doped materials is considered as a potential way to provide luminescence spanning in a wide wavelength range. Divalent and trivalent oxidation states of Eu provide emitting centers in the visible region. In consideration, the use of Eu in photonics requires Eu doped thin films that are compatible with CMOS technology but for example faces material science issues like a low Eu solid solubility in silica. Therefore approaches aim to obtain efficient light emission from silicon oxycarbide which has a luminescence in the visible range and can be a host material for rare earth ions. The silicon oxycarbide material can provide potential applications of the Eu luminescent materials to challenging conditions like high temperatures or aggressive environments where the silica has weaknesses. In some approaches, silicon rich silicon oxide that contain silicon nanoclusters emit red to near infrared luminescence due to quantum confinement effects while luminescence at shorter wavelength is difficult due to the interplay of defects and quantum confinement effects. In addition it is applicable as low-k dielectric, etch-stop and passivation layers. It also has an optical band-gap that is smaller than that of SiO2 which may facilitate carrier injection at lower voltages that is suitable for optoelectronics.

From materials perspective of emerging materials, it seems distant to consider system related issues. The future demands on communication and lighting devices require higher information flows in modernized optical devices, for example by replacing electrical interconnects with their optical counterparts and tunable backgrounds filters for integrated optics or photonics applications. However, there are materials issues related to such device performance, for example by a non-linearity, that provide the possibility for selective removal or addition of wavelengths using hetero structures in which one side of the structure enhances the light-to-dark sensitivity of long and medium wavelength channels and diminish others, and an opposite behavior in other face of the structure. Certainly materials may be applied in various innovative ways to provide new performances in devices and systems.

In any materials and device evaluation, reliability issues in passivation and packaging of semiconductor device structures provide a base knowledge that may be used to evaluate new concepts. Fundamental aspects of dielectric constant, bandgap and band offsets between the valence and conduction band edges between the passivation layer and the semiconductor create a foundation for understanding the device performance. In relation to these, the surface pre-treatment and deposition technique can influence the reliability and electric field durability of the system, and relate to interface and near interface regions between the dielectric and semiconductor which can host electronic defects which change the surface potential, reduces mobility and enhance the recombination of charge carriers.

At the end, materials for energy savings are critically needed. At the symposium "Alternative approaches of SiC and related wide bandgap materials in light emitting and solar cell applications", held at the E-MRS 2013 Spring meeting, 27–31 May , 2013 Strasbourg, France, a variety of concepts were presented. In this publication, a selection is presented that represents a range of issues from materials to reliability processing to system approaches.

Acknowledgements: Technical support during preparation of the symposium program and proceedings by Saskia Schimmel is greatly acknowledged.

Peer review statement

All papers published in this volume of *IOP Conference Series: Materials Science and Engineering* have been peer reviewed through processes administered by the proceedings Editors. Reviews were conducted by expert referees to the professional and scientific standards expected of a proceedings journal published by IOP Publishing.

Nucleation and growth of polycrystalline SiC

M Kaiser[1], S Schimmel[1], V Jokubavicius[2], M K Linnarsson[3], H Ou[4], M Syväjärvi[2] and P Wellmann[1]

[1]Materials Department 6, University of Erlangen, Martensstrasse 7, D-91058 Erlangen, Germany
[2]Department of Physics, Chemistry and Biology, Linköping University, SE-58183 Linköping, Sweden
[3]School of Information and Communication Technology, KTH Royal Institute of Technology, SE-16440 Kista, Sweden
[4]Department of Photonics Engineering, Technical University of Denmark, DK-2800 Lyngby, Denmark

E-mail: michl.kaiser@ww.uni-erlangen.de

Abstract. The nucleation and bulk growth of polycrystalline SiC in a 2 inch PVT setup using isostatic and pyrolytic graphite as substrates was studied. Textured nucleation occurs under near-thermal equilibrium conditions at the initial growth stage with hexagonal platelet shaped crystallites of 4H, 6H and 15R polytypes. It is found that pyrolytic graphite results in enhanced texturing of the nucleating gas species. Reducing the pressure leads to growth of the crystallites until a closed polycrystalline SiC layer containing voids with a rough surface is developed. Bulk growth was conducted at 35 mbar Ar pressure at 2250°C in diffusion limited mass transport regime generating a convex shaped growth form of the solid-gas interface leading to lateral expansion of virtually [001] oriented crystallites. Growth at 2350°C led to the stabilization of 6H polytypic grains. The micropipe density in the bulk strongly depends on the substrate used.

1. Introduction

Fluorescent silicon carbide (f-SiC) is a promising candidate in the field of optoelectronics for the fabrication of a novel monolithic all-semiconductor white LED structure [1]. Such devices require thick layers (> 100μm) of high crystalline quality and proper doping. The *fast sublimation growth process* (FSGP), a sublimation epitaxial technique, is proven to fulfil these requirements for growing f-SiC [2] when adequately doped solid state source materials are applied. In the case of fluorescent SiC, the research is very new and donor-acceptor co-doped sources are necessary. As no such materials exist we have previously reported on polycrystalline nitrogen and boron doped bulk SiC crystals grown using the *physical vapour transport* (PVT) technique [3]. It was also stated elsewhere that such source material is beneficial regarding minimization of graphitisation in the source and achievable layer quality at high growth rates [4].

In this work we investigate the growth of SiC crystallites on isostatically pressed highly purified and hereon deposited pyrolytic graphite substrates at the initial stages of growth of polycrystalline bulk SiC source material, i.e. when diffusion limited mass transport is activated due to the decrease of the environmental argon system pressure. The shape of the crystals, their orientation and density distribution in radial direction of the substrate is treated using optical and electron microscopy (SEM) as well as x-ray powder diffraction (XRD) as analytical methods. The results are in agreement with the temperature field distribution inside the growth chamber derived from numerical calculations using the Virtual Reactor® code. Moreover, results from polycrystalline bulk growth experiments on the before mentioned substrates at higher temperature are presented and the influence on polytype stabilization and structural quality is discussed. Additionally, the modification of texture during bulk growth is explained in terms of growth kinetics.

IOP Conf. Series: Materials Science and Engineering **56** (2014) 012001 doi:10.1088/1757-899X/56/1/012001

2. Experimental

Polycrystalline SiC boules were grown by the physical vapour transport (PVT) technique [5] in an inductively heated 2 inch setup. The complete process consists of the following steps: (I) evacuation of the reactor and heating with moderate power, (II) filling to 800 mbar with Ar and heating untill the set temperature is reached, (III) pumping down to growth pressure and growth and (IV) refilling with Ar to 800 mbar and reduction of the heating power. Two kinds of substrates were used: (a) isostatically pressed highly purified graphite (referred to as *isostatic graphite*) and (b) isostatic graphite with a pyrolytic graphite coating (referred to as *pyrolytic graphite*). In Table 1 the characteristics of the two substrate materials are summarized. The main difference is that the pyrolytic graphite has no open porosity and is [001] textured [6] whereas the isostatic graphite is isotropic regarding its grain orientation. Isostatic graphite substrates were used as-machined resulting in relatively high surface roughness RMS values. Due to the pyrolytic coating the surface roughness is slightly reduced.

Table 1. Characteristics of the seeds used in this work

	Surface roughness (μm)	Total / open porosity (%)	Avarage grain size (μm)	Texture
Isostatic graphite	2.25 ± 0.16	18,6 / 10	7	No texture
Pyrolytic graphite	1.96 ± 0.49	< 7 / 0	-	[001]

In order to investigate the nucleation process and layer formation three interruption experiments were carried out (referred to as *sample A, B* and *C*) during the first part of process step (III) which is displayed in Figure 1. Sample A was ended up at 800 mbar after temperature reached convergence at 2185°C. Sample B was interrupted 90 min after pumping down was initialized at 266 mbar and 2195°C. Sample C was terminated 180 min after sample A when 35 mbar (growth pressure) was reached and the corresponding temperature was 2233°C. Note that pressure reduction follows an exponential decay and results in reduced thermal conductivity of the insulation which causes the temperature increase. Samples A and C were nucleated on isostatic graphite, sample B on pyrolytic graphite. Bulk growth experiments were conducted at 35 mbar Ar atmosphere with various doping concentrations for 50h growth time and are described in more detail elsewhere [3].

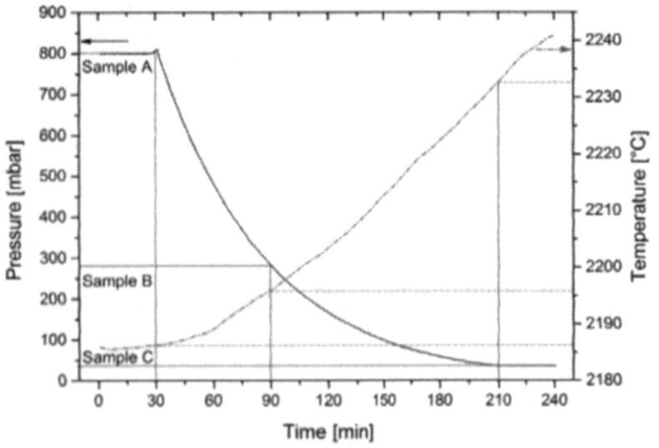

Figure 1. Excerpt of the growth process at the initial stage of nucleation showing the exponential decay of the pressure (solid line, left scale) and corresponding temperature (dash-dotted line, right scale). Interruption experiments A, B and C are denoted at the left side. Thin lines are drawn to guide the eyes.

Numerical modelling using the software package Virtual Reactor© was applied to determine and optimize the temperature field and distribution in the growth cell. The initial nuclei and grown layers were investigated using optical and scanning electron microscopy, micro-Raman spectroscopy as well as XRD in Bragg-Brentano geometry. Polycrystalline boules were cross-sectional sliced and polished with diamond suspension to surface roughness RMS values below 3 nm.

3. Results and Discussion

3.1. Nucleation of SiC on graphite

Sample A was aborted after the set temperature (2180°C) was reached at 800mbar. At this pressure regime in conjunction with the distance of 30 mm between the SiC powder source material and the graphite substrate, mass transport is expected to be low [7]. Moreover the system is in near-equilibrium conditions due to slow heating afore and the dwell period of 1h after the last heating step in which no significant temperature increase is measured by the optical pyrometer on top of the crucible. This assumption is confirmed by numerical simulation (Fig. 2 bottom) where the axial temperature gradient is calculated to be as low as 20 K/cm at the symmetry axis of the setup, i.e. the left border of the plot. Herein the temperature difference between isolines is 1 K. The calculated mass transport is displayed by small arrows where thick ones indicate increased transport compared to thin ones, corresponding to region I, III and II, respectively. Note that those regions are centro-symmetric circles when viewed normal to substrate surface which can be seen in the middle of Fig. 2 where a corresponding section of a photograph of the graphite substrate with nucleated crystals is shown. It is obvious that in region I and III the density of crystals is higher and the crystal size is increased compared to region II. Thus experiment and numerical modelling are consistent with each other.

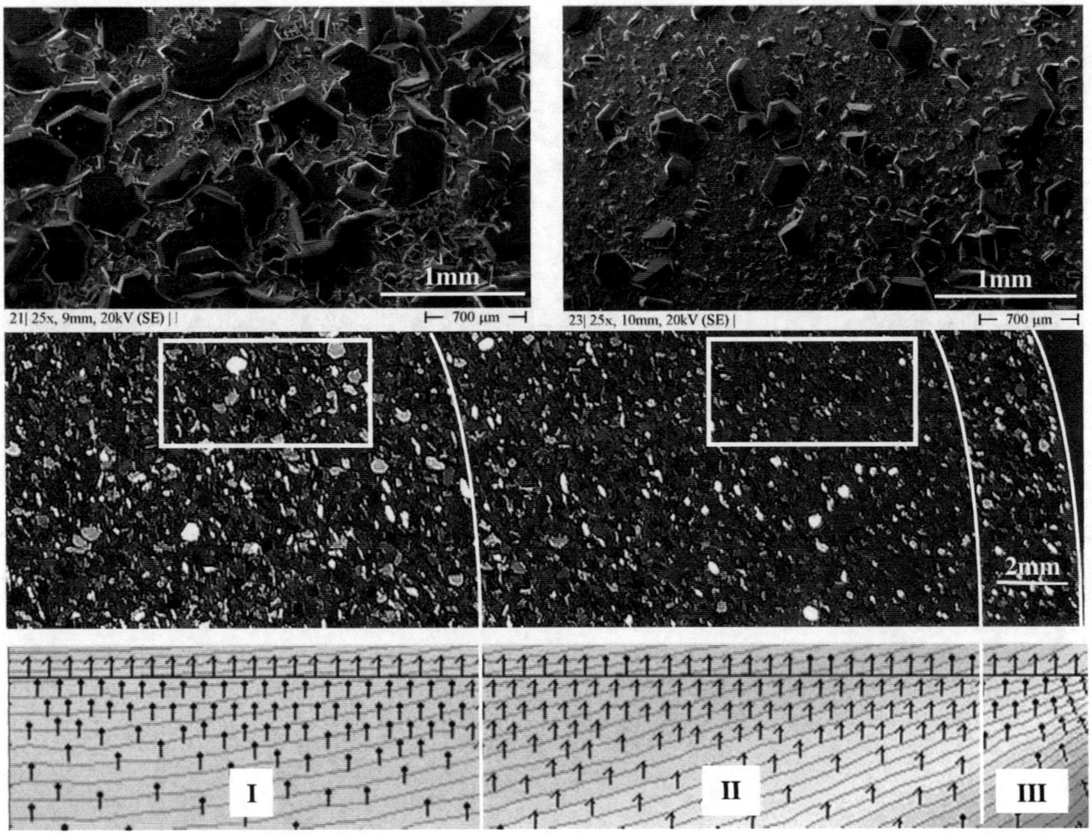

Figure 2. Numerical simulation of the temperature distribution near the graphite seed (bottom) for process parameters of sample A, respective section of a photograph (middle) and SEM images of the marked areas (top left and right). Temperature difference between isolines is 1K.

The reason for increased growth of crystals in region I and III is the lower temperature of the substrate and the increased axial temperature gradient, respectively, both resulting in higher local supersaturation compared to region II. Generally, when the supersaturation increases the nucleation frequency increases while the crystal size is supposed to decrease. The reason for the observed larger crystals in region I and III is that the initial nucleation occurred before sample A was aborted between si-rich gaseous species and the graphite. Thus the herein displayed crystallites result from grain selection and growth due to kinetics. Nevertheless, nucleation started earlier in the two regions mentioned before resulting in an increased growth time. A more detailed view into region I and II (Fig. 2 top) exhibits SiC crystals of hexagonal platelet shape, known as the close to the equilibrium shape of the 4H, 6H and the rhombohedral 15R polytype, deduced from spontaneous nucleation in Lely growth experiments. Such crystals reveal large {001} faces as well as inclined {h0l} pyramidal and {100} prismatic planes. As can be seen the majority of crystals show inclination between the <001> direction and the normal to the surface (Fig. 3 top left). Moreover, growth occurs dominantly along the <110> directions on the {001} faces resulting in the formation of the above mentioned pyramidal and prismatic planes and the platelet nature of the crystals. Note that no crystals were found showing the kinetic form of the cubic polytype which was also confirmed by micro-Raman measurements. That is because the process conditions are not favourable for the growth of β-SiC that is known to appear at lower temperatures and high temperature gradients resulting in high supersaturations.

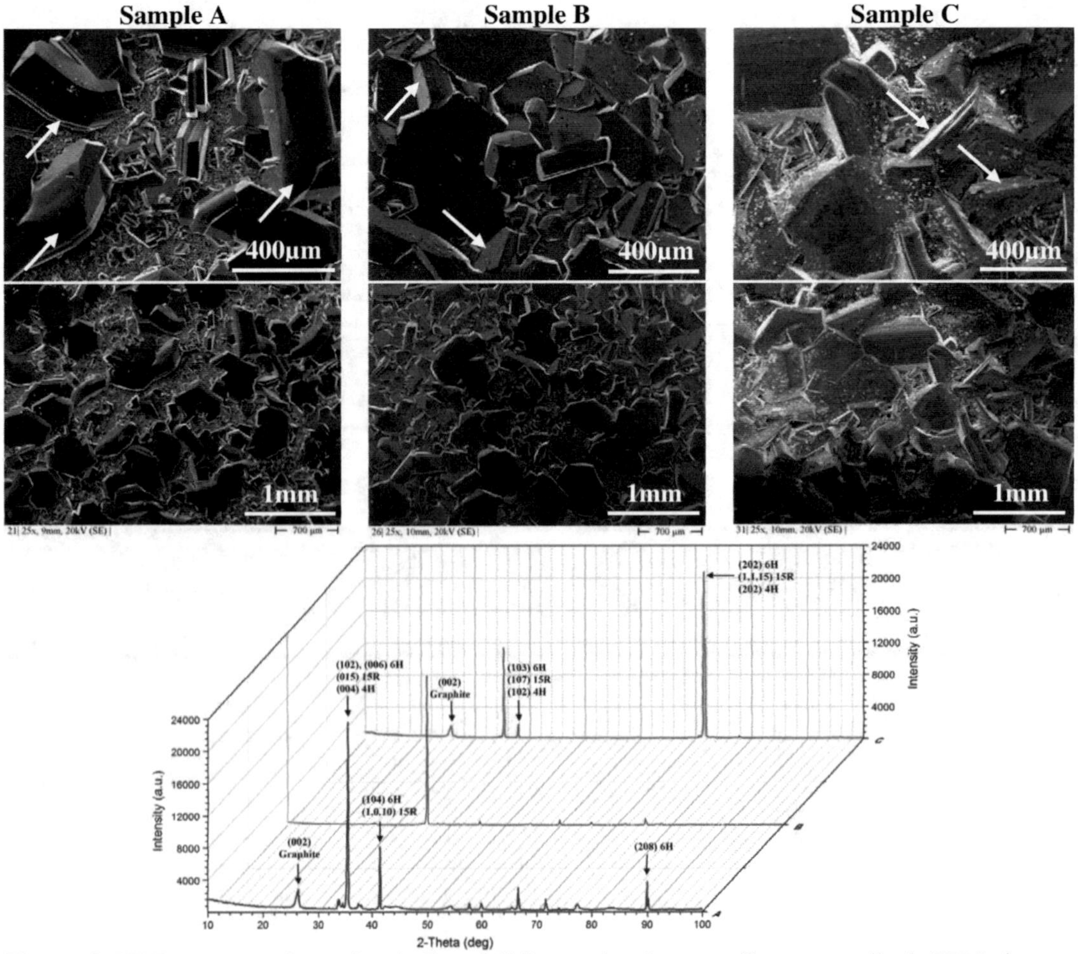

Figure 3. XRD patterns of samples A, B and C (bottom), relevant reflexes are edited. SEM pictures of the samples (middle) and corresponding outtakes (top).

In order to investigate the further growth of the crystals and to determine when a closed poly-SiC layer is formed on the graphite substrate experiments B and C were carried out and XRD powder diffraction was measured (Fig. 3 bottom) where diffraction peaks were identified using powder diffraction files PDF 00-029-1131 for 6H, PDF 00-022-1317 for 4H and PDF 01-072-4530 for 15R. From the XRD and Raman measurements it is derived that the crystals are a mixture of the 4H, 6H and 15R polytype, which can also be seen from the cross-sectional cut of a bulk crystal in Fig. 4. When taking into account relative diffraction intensities from homo-polytypic SiC powders with statistically distributed crystals one can conclude that texture is present at all samples, nevertheless it is hardly possible to distinguish to which amount each polytype incurs into the total intensity of the individual diffraction peaks since multiple reflexes may overlap. Despite this, the (102), (202) planes of 6H and the (015), (1,1,15) planes of 15R do incline approximately 62°, 75° and 71°, 73° with the {001} face, respectively, and such crystals can be seen on the corresponding SEM images (see arrows in Fig. 3 top left). Sample B was grown on pyrolytic graphite and the XRD pattern shows one distinct reflex at 2-Theta 35.7° accounting for growth on (102) and the basal plane of 6H, since lattice spacings are nearly equal. Moreover, the reflex also accounts for (105) of 15R as well as <001> oriented 4H. Nevertheless, micro-Raman measurements revealed that the majority of crystallites are of 6H polytype. Thus growth on pyrolytic graphite results in enhanced texturing but not solely on the {001} faces (Fig. 3 middle column). Regarding the coverage of the graphite substrate surface it can be mentioned that a closed layer containing voids is present at the process conditions of sample II. Such voids never completely disappear during on-going growth and are the reason for the formation of micropipes penetrating the bulk crystal as described to some more extend in the following section.

3.2. Bulk growth of polycrystalline SiC

Bulk crystals were grown at 35 mbar Ar pressure for 50h growth time at temperatures of 2250° or 2350°C on isostatic and pyrolytic graphite substrates. Since at this pressure diffusion limited mass transport is present, growth forms of the crystals are observed displaying the curvature of the temperature field inside the crucible (Fig. 4). Contrary to the close to the equilibrium shapes of the crystallites during the initial growth stages resulting in a rough surface, the latter begins to smoothen with shapes that are convex. Note that mass transport in Fig. 2 is calculated for the initial growth stages at 800 mbar and depicts only the region close to the graphite gas interface therefor most of the mass flow vectors are aligned normal to the substrate surface. The resulting growth form (calculation not shown here) develops during growth in which grain selection and coarsening commences. It was shown elsewhere that grains with low inclination angles between the [001] direction and the normal to the surface laterally overgrow heavily inclined grains resulting in less variation of the off-axis orientation of the individual grains during growth, i.e. texturing towards [001] is observed [3,8]. This lateral overgrowth can lead to the formation of voids in areas where heavily inclined platelets enclose empty space that is not filled before being overgrown. Thus a micro-cavity is formed in which a temperature gradient is present leading to sublimation within the bulk material and the formation of micropipes propagating along the growth direction.

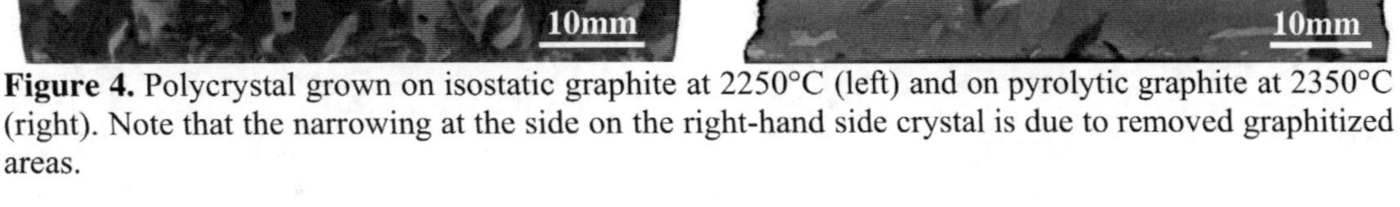

Figure 4. Polycrystal grown on isostatic graphite at 2250°C (left) and on pyrolytic graphite at 2350°C (right). Note that the narrowing at the side on the right-hand side crystal is due to removed graphitized areas.

It was found that the adoption of pyrolytic graphite strongly reduces the micropipe formation which is attributed to the fact that the initially grown crystallites are highly textured and thereby the probability of empty space formation between individual platelets is reduced. Bulk growth at 2350°C led to the stabilization of the 6H polytype for most of the grains (Fig. 4 right) which is known to be the high-temperature stable modification of SiC during bulk growth. Compared to growth at 2250°C, where grains in the millimetre range are obtained, grains up to 20 mm in length were established depicting the fact that the grain size as well as the polytype is a function of the growth temperature.

4. Conclusion

We have grown polycrystalline SiC bulk crystals on either isostatic or pyrolytic graphite substrates at 2250°C and 2350°C in a 2 inch PVT growth setup at 35mbar Ar pressure. Growth of initial crystallites of SiC was investigated at 800 mbar, 266 mbar and when reaching 35 mbar. It was found that under near-thermal equilibrium conditions at 800 mbar crystals exhibit forms close to the equilibrium shape. Their radial density distribution and size is a function of the temperature field and the corresponding temperature gradients inside the growth cell. When employing pyrolytic graphite as substrate increased texturing towards the [001] direction is observed. When mass transport is initiated the crystals emerge and form a dense polycrystalline layer with a rough surface and included voids being the reason for micropipe formation in the bulk, which can be significantly suppressed by using pyrolytic graphite. Due to the formation of growth forms the surface smoothens and lateral overgrowth takes place leading to grain selection and coarsening. Bulk growth at 2350°C led to the preferential formation of 6H polytypic grains.

Acknowledgement

We thank Marco Brandl from the Chair for Crystallography and Structural Physics, University of Erlangen for carrying out XRD measurements and fruitful discussion. Financial support by the STAEDTLER Foundation (contract SH/eh 17/12) is gratefully acknowledged.

References

[1] Kamiyama S, Iwaya M, Takeuchi T, Akasaki I, Syväjärvi M, Yakimova R 2011 *J. Semicond.* **32** 013004

[2] Syväjärvi M, Yakimova R, Jacobsson H, Janzén E 2000 *J.Appl. Phys.* **88** 1407

[3] Kaiser M, Hupfer T, Jokubavicus V, Schimmel S, Syväjärvi M, Ou Y, Ou H, Linnarsson M K, Wellmann P 2013 *Mat. Sci. Forum* **740** 39

[4] Jokubavicius V, Hens P, Liljedahl R, Sun J W, Kaiser M, Wellmann P, Sano S, Yakimova R, Kamiyama S, Syväjärvi M 2012 *Thin Solid Films* **552** 7

[5] Wellmann P, Desperrier P,Müller R, Straubinger T, Winnacker A, Baillet F, Blanquet E, Dedulle J M, Pons M 2005 J. Cryst. Growth 275 e555

[6] Bourrat X, Trouvat B, Limousin G, Vignoles G, Doux F 2000 *J. Mat. Res.* **15** 92

[7] Chen Q S, Zhanga H, Prasada V, Balkasb C M, Yushinb N K, Wang S 2001 *J. Cryst. Growth* **224** 101

[8] Chichignoud G, Pons M, Blanquet E, Chaussende D, Anikin M, Pernot E, Mermoux M, Madar R, Moisson C, Letertre F 2006 *Surf. Coat. Tech.***201** 4014

IOP Conf. Series: Materials Science and Engineering **56** (2014) 012002 doi:10.1088/1757-899X/56/1/012002

The role of defects in fluorescent silicon carbide layers grown by sublimation epitaxy

S Schimmel[1], M Kaiser[1], V Jokubavicius[2], Y Ou[3], P Hens[2], M K Linnarsson[4], J Sun[2], R Liljedahl[2], H Ou[3], M Syväjärvi[2], P Wellmann[1]

[1]Materials Department 6, University of Erlangen, Martensstrasse 7, D-91058 Erlangen, Germany
[2]Department of Physics, Chemistry and Biology, Linköping University, SE-58183 Linköping, Sweden
[3]Department of Photonics Engineering, Technical University of Denmark, DK-2800 Lyngby, Denmark
[4]School of Information and Communication Technology, KTH Royal Institute of Technology, SE-16440 Kista, Sweden

E-mail: saskia.schimmel@ww.uni-erlangen.de

Abstract. Donor-acceptor co-doped SiC is a promising light converter for novel monolithic all-semiconductor white LEDs due to its broad-band donor-acceptor pair luminescence and potentially high internal quantum efficiency. Besides sufficiently high doping concentrations in an appropriate ratio yielding short radiative lifetimes, long nonradiative lifetimes are crucial for efficient light conversion. The impact of different types of defects is studied by characterizing fluorescent silicon carbide layers with regard to photoluminescence intensity, homogeneity and efficiency taking into account dislocation density and distribution. Different doping concentrations and variations in gas phase composition and pressure are investigated.

1. Introduction

Donor-acceptor co-doped SiC was first recognized as a potential light converter for a novel monolithic all-semiconductor white LED by Kamiyama et al. in 2006 [1]. Since then, considerable research efforts have been made. However, the internal quantum efficiency (IQE) of typical samples lags far behind the theoretically predicted high efficiencies, even though a large range of doping concentrations and ratios has been studied. This suggests that the IQE is severely affected by short nonradiative lifetimes.

The layers are commonly grown by the Fast Sublimation Growth Process (FSGP) [2] which is highly suitable for the growth of high quality thick epitaxial layers with reasonable growth rate. According to XRD measurements, the epitaxial layers grown by FSGP may even show an improved crystalline quality with regard to the substrate [2] and the charge carrier lifetime is usually quite high, typically about 2 μs [3]. As the considerations above indicate that short nonradiative lifetimes may be a critical issue for the light conversion efficiency of f-SiC layers, this work will focus on the role of different types of defects in fluorescent silicon carbide.

2. Experimental

2.1. Growth of N-B co-doped 6H-SiC layers by sublimation epitaxy

Fluorescent silicon carbide layers were grown by the FSGP with 30 min growth time and 1850 °C growth temperature on 1.4 degree off-axis 6H-SiC substrates, yielding a layer thickness of 85 ± 10 μm. A polycrystalline monolithic source co-doped with nitrogen and boron was employed.

The source material was grown by physical vapour transport (PVT) and is described more in detail elsewhere [4].

Using the FSGP, both nitrogen and boron can be introduced by applying a co-doped monolithic silicon carbide source while nitrogen can alternatively or additionally be introduced via the ambient. Different process conditions, namely gas phase compositions and pressures (hereafter referred to as *growth ambient*) were applied as follows. Prior to each growth run, the reactor was evacuated for at least 15 h to remove adsorbed gases from the crucible and graphite insulation. In the conventional process, which will be denoted as *static RT*, the valve to the vacuum pump is closed at room temperature before starting temperature ramp-up. For nitrogen doping via the ambient, the reactor was filled with nitrogen after evacuation and pumped-down to the desired pressure. This will be denoted as nitrogen ambient accompanied by the actual pressure value. Note that there are two different pressure values: the pressure set at room temperature, and the pressure measured when reaching the growth temperature.

In order to further reduce the amount of gases adsorbed at the graphite parts of the reactor compared to static RT conditions, two new variants of static vacuum were tested. In contrast to static RT, pumping is continued during temperature ramp-up until a certain temperature is reached, namely either 1000 °C or 1250 °C. These two process conditions will be denoted as *static 1000 °C* and *static 1250 °C*, respectively.

Besides the different process conditions described above, the concentrations of the dopants and their ratio in the source material were varied. Process conditions and source doping for the series of samples used in this work are summarized in Table 1. As determined by Secondary Ion Mass Spectrometry (SIMS) measurements, a nitrogen flux of 8 sccm during source growth yields a nitrogen concentration of 1.5×10^{19} cm^{-3} ± 2.3×10^{18} cm^{-3} in the source material. 10 ppm, 50 ppm and 100 ppm B in the PVT source powder yield a source material boron concentration of 3.37×10^{17} cm^{-3} ± 5.1×10^{16} cm^{-3}, 1.47×10^{18} cm^{-3} ± 2.2×10^{17} cm^{-3}, and 3.00×10^{18} cm^{-3} ± 4.5×10^{17} cm^{-3}, respectively.

Table 1. Experimental conditions of series A. The given concentration of B corresponds to the amount of B_4C in the SiC source powder used in source growth by PVT while the amount of N is the nitrogen flow flux during source growth.

Sample (series A)	**335**	**336**	**337**	**338**	**339**	**340**	**341**	**342**	**343**
B [ppm]	50	50	50	50	50	50	50	50	50
N [sccm]	4	2	1	4	2	1	4	2	1
Ambient	Static RT	Static RT	Static RT	5×10^{-4} N_2	5×10^{-4} N_2	5×10^{-4} N_2	5×10^{-3} N_2	5×10^{-3} N_2	5×10^{-3} N_2
Sample (series B)	**374**	**375**	**376**	**377**	**378**	**379**	**386**	**387**	**388**
B [ppm]	10	10	50	50	50	50	100	100	100
N [sccm]	8	8	8	8	4	4	8	8	8
Ambient	Static RT	Static 1000 °C	Static RT	Static 1000 °C	Static RT	Static 1000 °C	Static RT	Static 1000 °C	Static 1250 °C

2.2. *Characterization of fluorescent 6H-SiC layers*

The f-SiC layers were characterized with regard to their photoluminescence intensity using room-temperature photoluminescence (PL) for series A and fluorospectrometry for series B.

In case of series A, the internal quantum efficiencies were determined. A half integration sphere was used for IQE determination with a 355 nm excitation laser yielding an excitation power density of 1.27 W/cm^2 (hereafter termed low excitation power density) and 25.5 W/cm^2 (hereafter termed high excitation power density), respectively. An estimated light extraction efficiency of 5 % was used for deducing IQE from the measurement. Charge carrier lifetimes where determined by microwave detected photoconductivity.

Series B was additionally characterized using room-temperature photoluminescence maps for studying luminescence homogeneity. After recording the PL maps, defect-selective etching was performed on one half of each sample of series B by dipping it into a KOH melt at 530 °C for 5 min. The etched samples were investigated using optical microscopy with regard to dislocation density.

3. Results and Discussion

In the following, photoluminescence and IQE measurements will be discussed with respect to the influence of defects on the internal quantum efficiency of f-SiC. Note that IQE of donor-acceptor pair luminescence depends on both radiative lifetime τ_r and nonradiative lifetime τ_{nr} as follows [3]:

$$\mathrm{IQE} = \frac{1}{1+\frac{\tau_r}{\tau_{nr}}} \qquad (1)$$

The radiative lifetime is determined by doping concentrations whereas the nonradiative lifetime depends on the concentration of defect energy levels providing nonradiative recombination paths. Thus, it is necessary to consider both doping and defect concentrations. Estimates of optimum doping concentrations and appropriate ratio of donors and acceptors in f-SiC are detailed elsewhere [5].

3.1. Photoluminescence maps and defect selective etching

Room-temperature photoluminescence maps in conjunction with optical microscopy maps of the sample after defect-selective etching show that a correlation of decreased luminescence intensity with dislocation density is observed only in areas of extremely high dislocation density. No difference in PL intensity is observed e.g. for the two insets A and B in Fig. 1 and 2 with dislocation densities of 15.4 x 10^4 cm^{-2} and 6.2 x 10^4 cm^{-2}, respectively. This indicates that the nonradiative lifetime is not dominated by dislocation density in most areas of the sample. This observation supports the conjecture that the nonradiative carrier lifetime in f-SiC is determined primarily by intrinsic defects or defect complexes.

3.2. The influence of doping concentrations and growth ambient on light conversion efficiency

For each of the source doping concentration used, the same trend was observed when varying the growth ambient (Fig. 3 and 4): All samples grown applying static 1000 °C show higher photoluminescence intensities than the corresponding static RT reference sample. Static 1250 °C yielded even higher luminescence intensity. This indicates that the adsorbed gases at the graphite parts cause some kind of defect that decreases the nonradiative lifetime. Total charge carrier lifetimes of samples 335 (high source nitrogen content, low nitrogen pressure in FSGP) and 343 (low source nitrogen content, high nitrogen pressure in FSGP) are 6.5 µs and 4 µs, respectively. This also supports that (nonradiative) charge carrier lifetime decreases if nitrogen is added to the growth ambient.

Similar conclusions can be drawn from the IQE measurements of the 4 sccm samples (Fig. 5) grown using different ambient nitrogen pressures which also show a decrease in efficiency with increasing nitrogen pressure in FSGP. The generally lower IQE at high excitation density conditions is attributed to the fact that filled donor and acceptor states result in enhanced nonradiative recombination. In contrast to the 4 sccm samples, the samples grown with 1 sccm and 2 sccm, respectively, do not show a clear tendency of IQE. This is most likely because the nitrogen ambient generally affects both radiative and nonradiative lifetime due to the increase in nitrogen doping and defect concentration, respectively. At very low source nitrogen doping, the decrease of radiative lifetime compensates for the decrease of nonradiative lifetime to an observable extent.

IOP Conf. Series: Materials Science and Engineering 56 (2014) 012002 doi:10.1088/1757-899X/56/1/012002

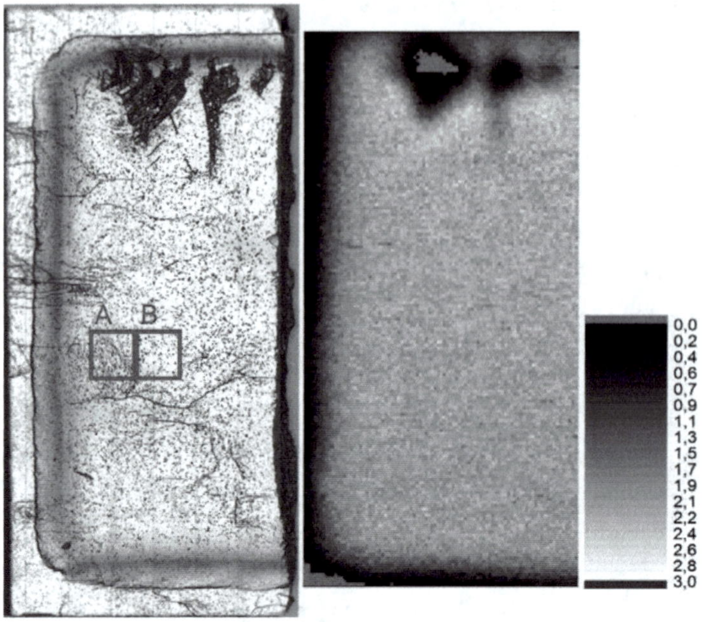

Fig. 1 Microscope map of sample ELS375 after defect-selective etching with the corresponding RTPL map (legend shows PL intensity in arbitrary units). The dark areas at the upper edge of the samples are inclusions of 3C-SiC.

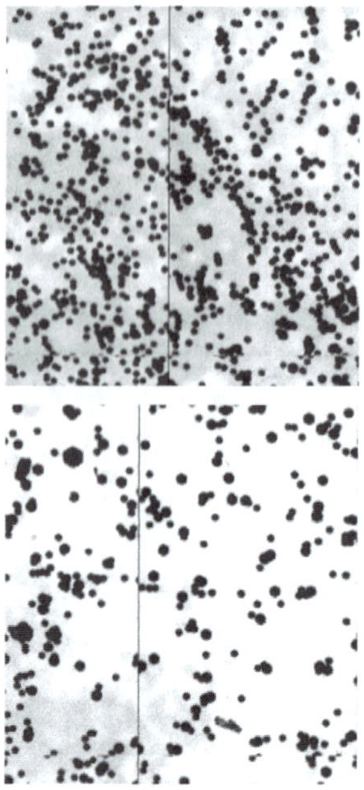

Fig. 2 Microscope images of the insets A (top) and B (bottom) marked in Fig. 5

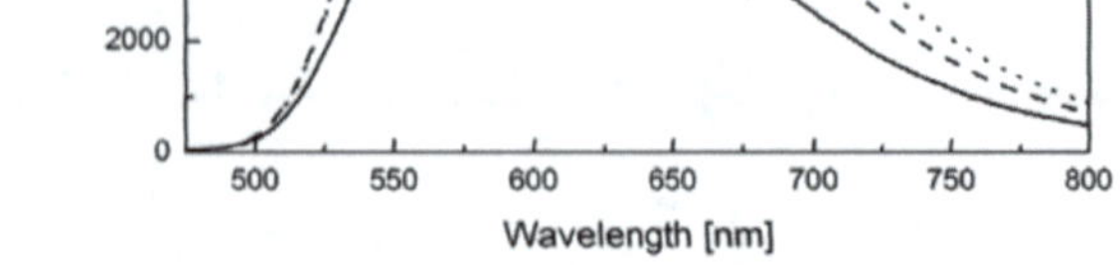

Fig. 3 PL spectra obtained by fluorospectrometry. The samples grown from 100 ppm B, 8 sccm N_2 are displayed as an example. Samples of different doping concentration show the same trend of increasing luminescence intensity if applying static RT and 1000 °C, as can be seen from the RTPL measurements shown in Fig. 4.

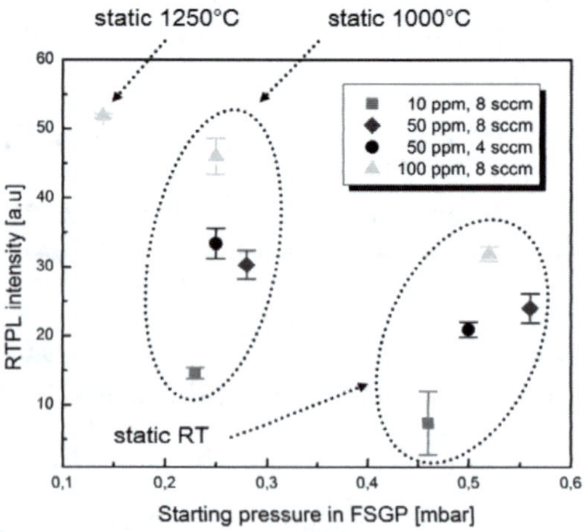

Fig. 4 Room temperature PL peak intensity as a function of starting pressure in sublimation epitaxy for samples grown from differently doped sources. Static vacuum modification used is indicated.

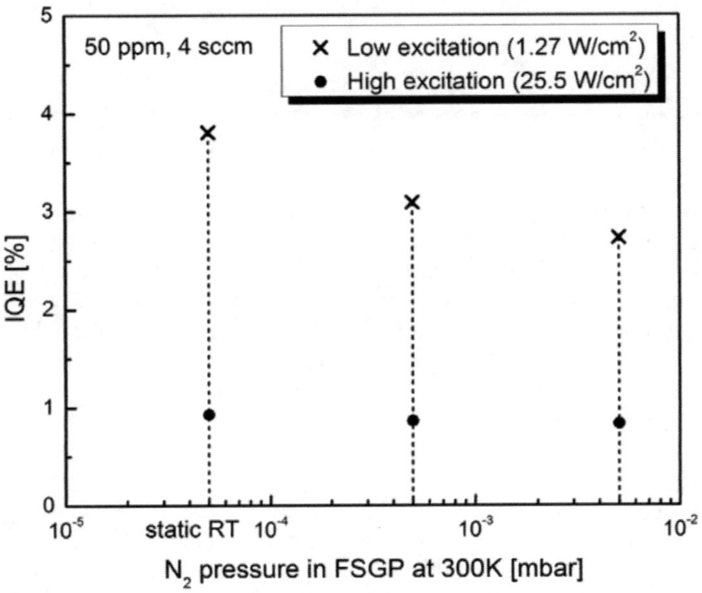

Fig. 5 Internal quantum efficiency of the 4 sccm samples of series A, for both low and high excitation.

The effect is not observed when increasing the pressure by introducing argon prior to growth, thus, it depends on the nature of the gas species and is not a general effect of slightly higher total pressure.

As the adsorbate is introduced from the air (when opening the reactor between growth runs), the possible influence of both nitrogen and oxygen needs to be considered. Experiments conducted with additional nitrogen pressure show a further decreasing luminescence intensity and internal quantum efficiency with increasing nitrogen pressure, indicating that nitrogen is likely to have the dominating influence. Interestingly it is not a matter of total nitrogen doping concentration in the epilayer, since the decrease in luminescence with increasing nitrogen pressure in the growth ambient is observed even at very low nitrogen doping concentrations in the source with an increase in source nitrogen doping still leading to considerably higher luminescence. Thus, nitrogen doping from the source (dispersed in atomic form within the source) and molecular nitrogen from the ambient must have a different effect on DAP luminescence efficiency, namely on the formation of defects that cause nonradiative recombination.

Considering that the bond dissociation energy of nitrogen is more than 50 times larger than the thermal energy at growth temperature, nitrogen can be assumed to be present primarily in molecular form in the growth ambient. Since the nature of the lifetime-limiting defect in f-SiC is not yet known for sure, one can only speculate on the mechanism of molecular nitrogen assisted defect formation. As common lifetime-limiting defects in silicon carbide are vacancy-related defects (such as the E_1E_2 in 6H-SiC and the $Z_{1/2}$ center in 4H-SiC) [6], one can suspect that adsorbed nitrogen molecules impede the incorporation of growth species into the lattice and hereby assist the formation of vacancy-related defects. The nitrogen molecule is expected to dissociate after adsorption since the energy of the N-N bond is significantly lowered by adsorption [7], however, it may be present as an adsorbed molecule for an average time sufficient to influence the incorporation of growth species, especially at high growth rates like in the FSGP. This assumption appears reasonable since nitrogen is also known to influence polytype stability [8]. Taking into account the model of nitrogen adsorption on the Si-face of (0001) 4H-SiC reported by Ashman et al. [9], it seems plausible that adsorbed nitrogen molecules may hinder the incorporation of carbon and hereby induce the formation of carbon vacancies or a defect complex containing a carbon vacancy.

4. Conclusion

The influence of growth ambient on the light conversion efficiency of N-B co-doped 6H-SiC layers grown by sublimation epitaxy was studied for samples grown from differently doped source material. The results confirm that high nitrogen and boron concentrations are needed for efficient photoluminescence, as discussed previously in [2]. However, adding nitrogen to the FSGP growth ambient results in a decrease of photoluminescence intensity and internal quantum efficiency for a broad range of source doping concentrations. Additionally, an effect of dislocation density on photoluminescence intensity is observed only in regions of exceptionally high dislocation density. These two findings strongly indicate that some kind of intrinsic defect or defect complex plays a dominating role in the internal quantum efficiency of fluorescent silicon carbide. The precise nature of the defect limiting the nonradiative lifetime is still to be determined and its nitrogen-assisted formation mechanism remains a matter of discussion.

Acknowledgements

Financial support by the German Ministry of Education and Research (BMBF, contract INNER 03SF0393), Ångpanneföreningen Research Foundation Swedish Energy Agency, Nordic Energy Research, Swedish Research Council (No. 2009-5307), Danish Council for Strategic research (No. 09-072118), and STAEDTLER Foundation (contract SH/eh 17/12) is gratefully acknowledged. One of the authors (S. S.) would like to thank the German Association for Crystal Growth (DGKK) and ERASMUS exchange program for a scholarship.

References

[1] Kamiyama S, Maeda T, Nakamura Y, Iwaya M, Amano H, Akasaki I, Kinoshita H, Furusho T, Yoshimoto M, Kimoto T, Suda J, Henry A, Ivanov I G, Bergman J P, Monemar B, Onuma T and Chichibu S F 2006 *J. Appl. Phys.* **99** 093108

[2] Syväyärvi M, Yakimova R, Tuominen M, Kakanakova-Georgieva A, MacMillan M, Henry A, Wahab Q and Janzén E 1999 *J. Crystal Growth* **197** 155–162

[3] Kamiyama S, Iwaya M, Takeuchi T, Akasaki I, Syväjärvi M and Yakimova R 2011 *J. Semicond.* **32(1)** 013004

[4] Kaiser M, Hupfer T, Jokubavicus V, Schimmel S, Syväjärvi M, Ou Y, Ou H, Linnarsson M K and Wellmann P 2013 *Mater. Sci. Forum* **740 – 742** 39-42

[5] Sun J W, Kamiyama S, Jokubavicius V, Peyre H, Yakimova R, Juillaguet S and Syväjärvi M 2012 *J. Phys. D: Appl. Phys.* **45** 235107

[6] Sasaki S, Kawahara K, Feng G, Alfieri G and Kimoto T 2011 *J. Appl. Phys.* **109** 013705

[7] Danielsson O, Forsberg U and Janzén E 2003 *J. Cryst. Growth* **250** 471-478

[8] Schmitt E, Straubinger T, Rasp M, Vogel M and Wohlfart A 2008 *J. Cryst. Growth* **310** 966-970

[9] Ashman C R and Pennington G 2008 *Surf. Science* **602** 3617-3622

Examination of Photoluminescence Temperature Dependencies in N-B Co-doped 6H-SiC

V Gavryushin[1], K Gulbinas[1], V Grivickas[1], M Karaliūnas[1], M Stasiūnas[1], V Jokubavičius[2], J W Sun[2], and M Syväjärvi[2]
[1]Institute of Applied Research, Vilnius University, Saulėtekio av. 10, Vilnius, LT-10223 Lithuania
[2]Department of Physics, Chemistry and Biology, Linköping University, Linköping, SE-58183 Sweden

E-mail: vytautas.grivickas@ff.vu.lt

Abstract. Two overlapping photoluminescent (PL) bands with a peaks (half-width) at 1.95 eV (0.45 eV) and 2.15 eV (0.25 eV), correspondingly at 300 K, are observed in heavily B-N co-doped 6H-SiC epilayers under high-level excitation condition. The low energy band dominates at low temperatures and decreases towards 300 K which is assigned to DAP emission from the nitrogen trap to the deep boron (dB) with phonon-assistance. The 2.15 eV band slightly increases with temperature and becomes comparable with the former one at 300 K. We present a modelling comprising electron de-trapping from the N-trap, i.e. calculating trapping and de-trapping probabilities. The T-dependence of the 2.15 eV band can be explained by free electron emission from the conduction band into the dB center provided by similar phonon-assistance.

1. Introduction

Recently it was demonstrated that boron-nitrogen (B-N) co-doped 6H-SiC epilayers exhibits efficient donor-acceptor pair (DAP) light emission in a broad green-red spectral range [1-3]. Combining it with a common nitride based UV excitation from grown III-V nitrides on the top of 6H-SiC, this PL emission opens ability for a proper light conversion using a semiconductor as phosphor. In addition, PL emitted by Al-N DAP gives a broad blue-green spectral range and by assembling these two layers, the combination produces a white light emitting SiC source with high colour rendering parameters at room temperature (RT) [2].

Still, the PL emission mechanism itself, and particularly at the high co-doping, is not well understood. Up to now, very little is known on the donor-acceptor clustering, the deep defect structure and/or its modification with doping and how all these factors may influence the PL emission. In this work, the epitaxial heavily co-doped B-N thick 6H-SiC layers have been grown by sublimation growth process. The photoluminescence was studied at high injections which allowed total filling of the shallow and deep traps. We concentrated our study on the temperature and injection dependencies of the photoluminescence spectra in order to separate the main radiative emission components.

2. Samples and experimental techniques

The fast sublimation growth process was used to grow homoepitaxial co-doped layers on low-off axis 6H-SiC substrates. N and B were introduced via the controlled polycrystalline SiC source and the N concentration was also enhanced by the nitrogen gas ambient during the growth. In this work, two

IOP Conf. Series: Materials Science and Engineering 56 (2014) 012003 doi:10.1088/1757-899X/56/1/012003

samples both having a long carrier lifetime (detected in high injection up to ms-time range) were investigated. The sample ELS118 had epi-thickness d = 45 μm, the incorporated concentration N (9.2×10^{18} cm^{-3}), and B (5.2×10^{18} cm^{-3}). The sample ELS296 had the corresponding values of d = 200 μm, N (1.2×10^{19} cm^{-3}), and B (2.1×10^{18} cm^{-3}). The samples were placed in a close-cycle cryostat in a cold finger configuration and were excited with 355 nm 25 ps light pulses from a frequency tripled YAG laser which was operating at 10 Hz repetition rate. The excitation light penetration depth is about 11 μm and is nearly independent of temperature. The time-integrated PL was collected by a lens in 90° geometry, focusing the emission onto the entrance slit of the Hamamatsu spectrograph, while light detection was achieved through the intensified CCD-array. The spectral resolution was about 0.8 nm.

3. Experimental results and analysis

In figure 1a we plot the measured PL spectra from 8 K to 300 K with an increasing step from 5 K to 20 K. The constant injection fluence of 1.5 mJ/cm^2 was maintained during these measurements. The PL in the sample ELS118 is about 3-5 times higher than that in the sample ELS296 (note the difference in scale). Both samples emit in a similar broad range in which the PL maximum is shifting gradually to the blue wavelength from 680-705 nm to 590-580 nm with increasing temperature. At

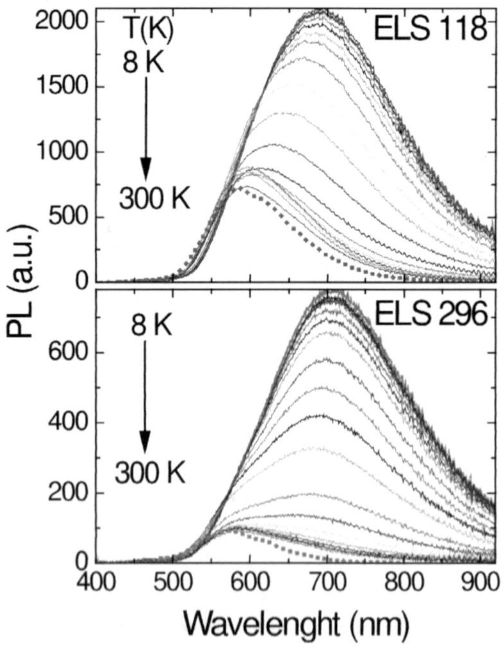

Figure 1a. PL spectra of two samples at different temperature. The excitation fluence is 1.5 mJ/cm^2.

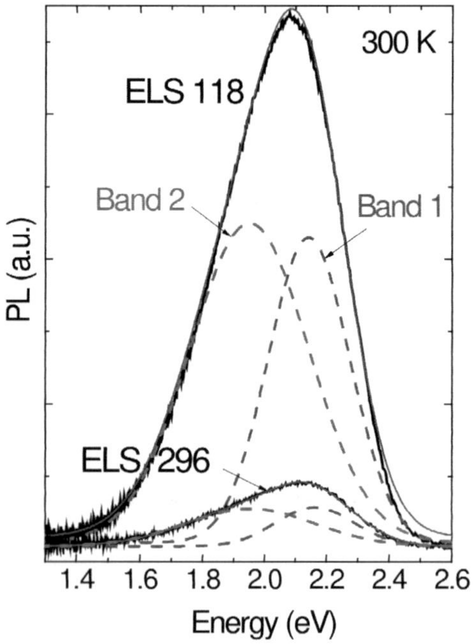

Figure 1b. Fitting of two Gaussian bands for PL spectrum of two samples at 300 K.

room temperature PL was measured also by an orders weaker cw-excitation, as reported in [2, 3]. The arbitrary normalized data are shown here by the dotted curves in figure 1a along with the high-injection data. Clearly, the high-injection RT PL data display substantial enlargement on the red emitted spectrum side. It is worth to note that for the cw-excitation a weaker shoulder was detected at 610-630 nm. Appearance of this shoulder has been attributed tentatively to the phonon-assisted replica in the N-dB DAP emission mechanism [2, 3]. For the high-injection conditions we did not record such shoulder at any injection fluence which was varied between 0.025-2.5 mJ/cm^2. This absence can be explained by a fact that the shoulder is over helmed by an intensity enlargement on the red wavelength emission side.

The obtained higher temperature PL spectra (particularly in the ELS296 sample in figure 1a), appear to have a composition of two overlapping bands with a quite different temperature behaviour. In order to explore this behaviour, we have performed a fitting procedure of each PL spectrum using a free parameter variation by two Gaussian shapes. The fitting examples (hereafter denoted as "Band 1" and "Band 2") are show in figure 1b at 300 K. Obviously the corresponding sum of the two bands, as given by the fine line, exhibits a good first-rate agreement to the experimental data in both samples. The similar fit was achieved at other temperatures as well.

In figure 2a the fitting parameters are shown, i.e. the peak energies (E), the full-with half-maxima (FWHM) and the band amplitudes as a function of temperature. Solid symbols belong to the ELS118 and the open ones to the ELS296. The peak energy (FWHM) of the "Band 1" is 2.15 eV (0.45 eV) and the corresponding value of the "Band 2" is ~1.9 eV (0.25 eV). Peak energies E for the "Band 2" clearly shift towards high energy from 1.8 eV to 1.95 eV with rising temperature while for the "Band 1" a smaller shift is observed only for ELS118. The FWHM of both bands is slightly shrinking with T rising to 300 K. As can be noted from the lowest panel of figure 2a, the "Band 1" amplitude steeply decreases with T increasing and quite opposite trend is observed for the "Band 2". Lastly, the amplitude increases with T for ELS118 or is about at a constant value for ELS296. In any case, the applicability of the two-band approach is appropriate above 100 K. At 300 K both bands are almost of the same amplitude. On the other hand, at T < 50 K, the "Band 2" alone can quite well represent the total PL spectrum. In figure 2b we show a single Gauss fit for the measured spectra at 8 K. The peak E is almost independent on the excitation fluence and the difference in the peak position is about 0.05 eV for both samples. However, the FWHM is slightly shrinking with increasing excitation fluencies. The PL amplitude increases linearly (a power factor 1.0) at low fluencies. At higher fluencies above 0.5 mJ/cm^2 it tends to saturate. More than an order lower emission was resolvable in the ELS118 sample spectrum at 8 K in the high energy range between 2.6-2.8 eV, i.e. closer to the band edge. Its intensity dependence is presented in the lower panel of figure 2b, which shows that this emission is saturated at 0.2 mJ/cm^2. PL emission in this range can be caused by non-intentional Al acceptor through Al-N DAP mechanism. To summarize the obtained results we derive that all available traps are perhaps under saturation at 1.5 mJ/cm^2. The estimated injected pair density at the absorption depth for this fluence is at about 10^{19} cm^{-3}.

Since the deep boron (dB) with energy ~550-750 meV is the dominant acceptor impurity in the investigated epilayer [1], the electron emission from nitrogen trap to the deep boron, i.e. DAP emission could be related to the "Band 2". On the other hand, the "Band 1" which intensity increases with T, can be assigned to the emission of free-electron transition to the deep boron. According to Thomas et al. [6], the energy of DAP recombination is given in first approximation by the equation: $E(r) = E_g - (E_a + E_d) + e^2/\varepsilon r$ where E_g is the band gap energy; E_a and E_d are the ionization energies of acceptor and donor; ε is the dielectric constant. Electron and hole pairs located in a close lattice sites present the strongest coulumb interaction and their recombination results in emission of larger energy photons contributing to the blue side of the DAP band [6]. The energy of phonons emitted by infinitely distant pairs is $E(\infty) = E_g - (E_a + E_d)$. Therefore, emission from DAP "Band 2", estimating $E_g = 2.95$ eV, $E_a = 0.65$ eV, $E_d = 0.1$ eV, would account for $E > 2.2$ eV. For the case of emission from conduction band-to-dB (for "Band 1") it should be set $E_d = 0$ eV which corresponds to emission at $E >$ 2.3 eV. Both approaches cannot account for the low energy side of observed PL bands. Thus, in the DAP recombination model it should be pointed out that either a large energy is taken by the phonons for the indirect bandgap SiC or the impurities are distributed in a wide energy range within the band gap. An alternative approch could be excitonic recombination mechanisms at the deep trap but this mechanism is unlikely at high-injection conditions.

In figure 3a the Arrhenius plot for the the amplitude of "Band 2" in both samples is shown. Excluding the high T range of the sample ELS118, we may estimate a trend from the PL slope that an activation energy of the trap is 53 meV. This value is in a perfect agreement to the same value deduced for the hundred-μs-long electron liftime component in the sample ELS118 [see Manolis et al. this issue].

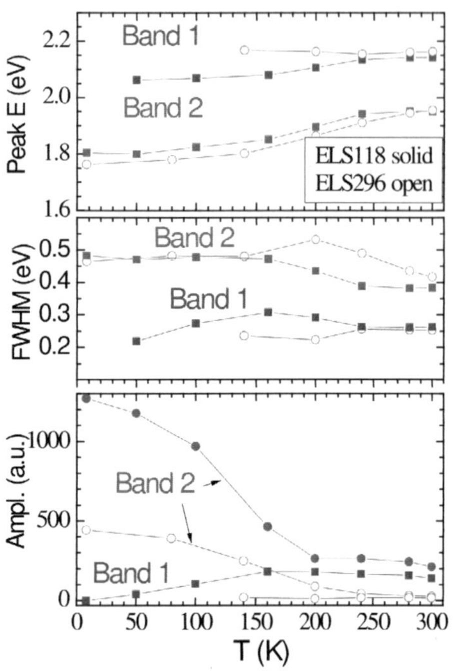

Figure 2a. T-dependencies of "Band 1" and "Band 2" peak energy, FWHM and the relative amplitudes at the excitation fluence of 1.5 mJ/cm^2.

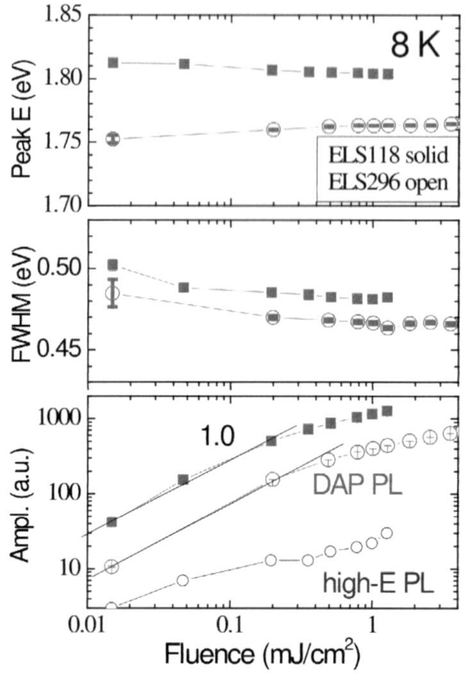

Figure 2b. The dependences on fluence of PL peak energy, FWHM and the amplitude in two samples at 8 K (a Gauss fit corresponds to the "Band 2").

Concerning the basic assumption for two band model, we provide a carrier population calculation. The calculation is based on a simple recombination model consisting of one trapping and one recombination center for the excess electrons. We solved a set of coupled rate equations for the free electrons $n(t,T)$, trapped electrons $n_t(t,T)$, and electrons captured by recombination centers $n_r(t,T)$, like in a model presented by Ichimura [7]. Thermal velocity, ionization rates and the effective density of states in the conduction band were taken with a 200-μs long recombination lifetime and the capture cross-section for the neutral-charged N-trap was fixed at the typical value of $\sigma_n^t = 2\ 10^{-17}$ cm^2. Then, we calculated carrier occupation in the N-trap and in the conduction band at quasi-thermodynamic equilibrium after the excitation pulse. Calculation was performed for trap levels E_t in the range from 10 meV to 100 meV at different temperatures. A plot in figure 3b shows two curve sets that are applied to the PL results for the "Band 2" and the "Band 1". The experimental points are taken from ELS118 sample of figure 2a. Carrier density in the N-trap (solid curves) was arbitrarily adjusted at the low temperature to the experimental PL amplitude value of 1280 (a.u.). A family of carrier densities in the conduction band (dashed curves) was adjusted to the N-trap value by multiplying the calculated result by the factor ×0.4, i.e. assuming slightly reduced emission probability. The dashed-dotted line in figure 3b shows the T-dependence in the case of no trap levels $N_t = 0$. In total, figure 3b plot reveals that occupation of N-trap and the conduction band can satisfactorily explain both PL band T-dependencies. The estimated thermal activation energy of N-trap is, however, lower, perhaps $E_t \approx$ 10-

20 meV. The substantial deviation of experimental point from the curves occurs at T > 150 K. This can be explained by the fact that radiation emission probability was taken as a constant value, without taking into account the increase of phonon density. In order to verify this hypothesis, a proper mechanism for phonon-assistance in PL emission mechanism should be developed.

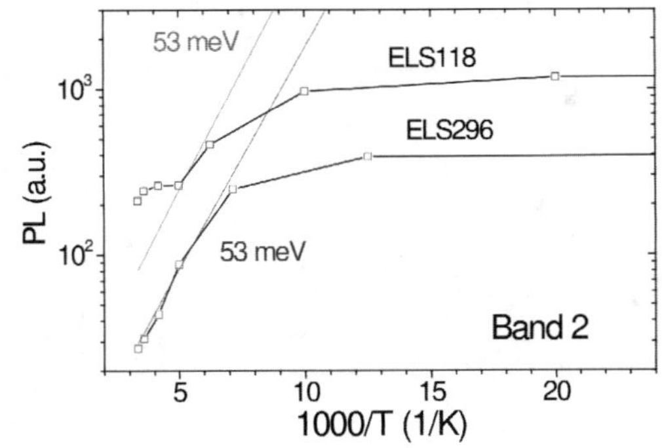

Figure 3a. Arrhenius plot of PL for the amplitude of the "Band 2".

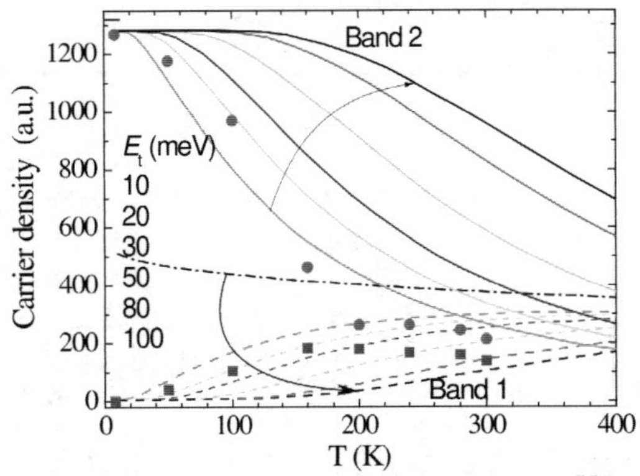

Figure 3b. T-dependencies of the electron density in the conduction band (Band 1) and in the N-trap (Band 2) at different trap energies. The experimental points are taken for ELS118 sample.

4. Conclusions

By applying PL measurements at high-injection levels we have shown that PL emission spectra are composed of two distinct broad bands "Band 1" and "Band 2" with a strictly different T-dependency. The RT PL data display substantial enlargement on the red emitted spectra side in respect to low cw-excitation spectra. This effect can be attributed to the enhanced phonon-assistance at high excitation. The low energy band we relate to the phonon-assisted N-dB DAP radiation with a thermal activation energy of 53 meV which is assigned for the N-trap. While the high energy band component is attributed to the phonon-assisted free-electron radiation recombination from the conduction band into the dB trap. The general behaviour of the T-dependencies in PL bands is basically governed by the occupation probability of the N-trap.

References

[1] Kamiyama S, Maeda T, Nakamura Y, Iwada M, Amano H, Akasaki I, Kinoshita H, Furusho T, Yoshimoto M, Kmoto T, Suda J, Henry A, Ivanov I G, Bergman J P, Monemar B, Onuma T and Chichibu S F 2006 *J. Appl. Phys.* **99** 093108
[2] Syväjärvi M, Müller J, Sun J W, Grivickas V, Ou Y, Jokubavicius V, Hens P, Kaisr M, Ariyawong K, Gulbinas K, Liljedahl R, Linnarsson M K, Kamiyama S, Wellmann P, Spiecker E and Ou H 2012 *Physica Scripta* **T148** 014002
[3] Sun J W, Jokubavicius V, Liljedahl R, Yakimova R, Juillaguet S, Camassel J, Kamiyama S and Syväjärvi M 2012 *Thin Solid Films* **522** 33-35
[4] Duijn-Arnold A V, Icoma T, Poluektov O G, Baranov P G, Mokhov E N and Schmidt J 1998 *Phys. Rev. B* **57** 1607-19
[5] Lebedev A 1999 *Semiconductors* **33** 107-30
[6] Thomas D G, Hopfield J J and Augustyniak W M 1965 *Phys. Rev.* **140** A202-220
[7] Ichimura M 2006 *Solid-State Electronics* **50** 1761-6

IOP Conf. Series: Materials Science and Engineering **56** (2014) 012004 doi:10.1088/1757-899X/56/1/012004

Carrier Lifetimes and Influence of In-Grown Defects in N-B Co-Doped 6H-SiC

V Grivickas[1], K Gulbinas[1], V Jokubavičius[2], J W Sun[2], M Karaliūnas[1], S Kamiyama[3], M Linnarsson[4], M Kaiser[5], P Wellmann[5] and M Syväjärvi[2]

[1]Institute of Applied Research, Vilnius University, Saulėtekio av. 10, Vilnius, 10223 Lithuania

[2]Department of Physics, Chemistry and Biology, Linköping University, Linköping, 58183 Sweden

[3]Faculty of Science and Technology, Meijo University, Tempaku-ku, Nagoya, 468-8502 Japan

[4]School of Information and Communication Technology, Royal Institute of Technology, Kista-Stockholm, 58183 Sweden

[5]Department of Materials Science – Materials for Electronics and Energy Technology, University of Erlangen-Nuremberg, Erlangen, 91058 Germany

E-mail: vytautas.grivickas@ff.vu.lt

Abstract. The thick N-B co-doped epilayers were grown by the fast sublimation growth method and the depth-resolved carrier lifetimes have been investigated by means of the free-carrier absorption (FCA) decay under perpendicular probe-pump measurement geometry. In some samples, we optically reveal in-grown carbon inclusions and polycrystalline defects of substantial concentration and show that these defects slow down excess carrier lifetime and prevent donor-acceptor pair photoluminescence (DAP PL). A pronounced electron lifetime reduction when injection level approaches the doping level was observed. It is caused by diffusion driven non-radiative recombination. However, the influence of surface recombination is small and insignificant at 300 K.

1. Introduction

B and N co-doped 6H-SiC epilayers combined with GaN-ultraviolet excitation chip can be used as an active material in a high-efficiency LED structure stack for a visible broad spectral light generation by DAP emission mechanism [1,2]. As accounted theoretically, this concept from blue excitation light could provide internal energy conversion up to 70% with a high color quality of a white spectrum. Moreover, in SiC it is possible to obtain uniform concentration of impurities in depth of hundred microns. It is well known that SiC material itself has excellent thermal conductivity property which is important for high power LED performance. Therefore, high co-doping in SiC requires further understanding of the growth techniques and investigation of optical properties. Up to now, very little is known on impurity clustering, on actual carrier transport mechanism and on the competing nonradiative recombination channels [3, 4]. In addition, an impact of co-doped acceptor situated on the hexagonal and the cubic sites in 6H-SiC matrix and their influence in the DAP processes is also not completely established yet [5].

In this work we present the free-carrier-absorption (FCA) lifetime studies performed on thick N-B doped epilayers in order to elucidate influence of surface and bulk recombination. We focus on previously observed pronounced injection dependence of the carrier lifetimes [3]. The surface recombination is estimated by performing the depth-resolved FCA measurement. Furthermore, we

IOP Conf. Series: Materials Science and Engineering **56** (2014) 012004 doi:10.1088/1757-899X/56/1/012004

examine overall carrier lifetime dependence on the concentration of in-grown threadlike defects and show that the only samples with low density of these defects exhibit DAP photoluminescence.

2. Experimental

The N-B doped 6H-SiC samples were grown by the fast sublimation growth method on low-off-axis 6H-SiC commercial substrates using N and B doped PVT crystalline SiC as a source material. The N

Table 1. Two types of doped 6H-SiC epilayers were studied (denoted ELS and RF batches).

Sample	ELS296	ELS297	RF18	RF04
Epilayer [μm]	200	200	130	110
B [cm^{-3}]	$2.1 \cdot 10^{18}$	$4 \cdot 10^{16}$	$2.5 \cdot 10^{18}$	$2.2 \cdot 10^{18}$
N [cm^{-3}]	$1.3 \cdot 10^{19}$	$1.1 \cdot 10^{19}$	$4 \cdot 10^{18}$	$8 \cdot 10^{18}$
Defects [cm^2]	10^{1} *a)	10^{6} *a)	10^{1}-10^{2} *b)	10^{5} *b)
Lifetime (μw-PC) [μs]	-	-	0.8	0.22
Lifetime component (FCA) [μs]	1.5 50	0.18	3 150 1200	0.1

*a) carbon inclusions; *b) polycrystalline inclusions

concentration was also enhanced by the N gas ambient during the growth [1]. Intentionally, the samples were grown to have higher donor concentration than acceptor concentration because the overcompensated *n*-type 6H-SiC provides more efficient DAP photoluminescence [1]. Estimated doping inhomogeneity varied about 20 - 30% along growth direction [see Gulbinas et al. in this issue]. Thickness of the epilayer was between 110 to 200 μm. The N and B concentrations, as determined by secondary ion mass spectrometry profiles, are provided in table 1. Unfortunately, it is not known how boron distributes between deep (dB) and shallow (sB) boron energy levels.

After growth the samples were cut into 0.5-0.7 mm-wide strips and lateral strips facets were polished to optical quality. The samples were inspected by Nomarsky microscope. The images of cross section view under polarized light showed in-grown defects with density varying widely from 10^1 cm^2 to 10^6 cm^2 (table 1). Side view revealed that thin filamentary defects were spreading out in groups along some straight line directions, like in figure 1a, and as a single defect, like in figure 1c. Top view, figure 1b, shows variety of directions along which the lines are able to spread. Closer inspection revealed that these filaments are composed of smaller dots. The alignment of the defects in the direction perpendicular to the interface implies that these defects are extended primarily along the threading dislocations. In the high defect density samples (RF-batch samples) it was found that the most of filamentary defects collapse into elliptical cones. Figures 1d, e show that those defects probably contain a polycrystalline SiC. Carbon concentration was determined by X-ray element analysis. It showed that these defects contain of 65-75 % carbon inside the cone. The cone defects appear visible on the surface top view as a phase inclusion of a 5-30 μm in size as shown in figure 1f. It is interesting to note that only in 10-20% of these cases cone defects are originating from the substrate/epilayer interface. In the rest 80-90% cases, they originate (or became visible) at a certain depth within the epilayer. This implies that suppression of defect growth might be achieved by a proper control of the growth conditions.

Technical details about the spatially and time-resolved FCA detection technique can be found in a review chapter [6]. For the carrier lifetime studies free carriers were excited by 2 ns laser pulses at 355 nm wavelength of the third harmonic from a YAG laser. The diameter of laser spot was about 1 mm^2, and the pulse repetition rate was equal to 40 Hz. If needed, the pump pulse fluence was reduced by optical filters. The induced FCA transient, $\Delta\alpha(t) = 1/d \cdot \ln(T_0/T(t))$, was detected at different depths by using 1.3 μm cw-probe beam which was directed perpendicular to the excited surface. The probing depth-resolution is set by the diameter of the probe beam within the crystal which was estimated to be about 7 μm. Using micrometer translation of the cryostat together with sample it was possible to choose the position of the probe beam with 1 μm precision at any distance from the excited surface. The transmitted probe beam was collected by lenses and optical transient was recorded by a fast photoreceiver with 0.5 ns time resolution. Then the probe decay was averaged by the digital oscilloscope over typically 500 traces in order to attain sufficient signal-to-noise ratio. The induced FCA transient at different depths was measured at the injection level varied from 10^{17} cm^{-3} to $8 \cdot 10^{18}$ cm^{-3}. Due to dominating absorption cross-section $\Delta\alpha(t)$ and a fast capture of minority hole to the deep traps, the decay mainly reflects the free excess electron absorption in the conduction band.

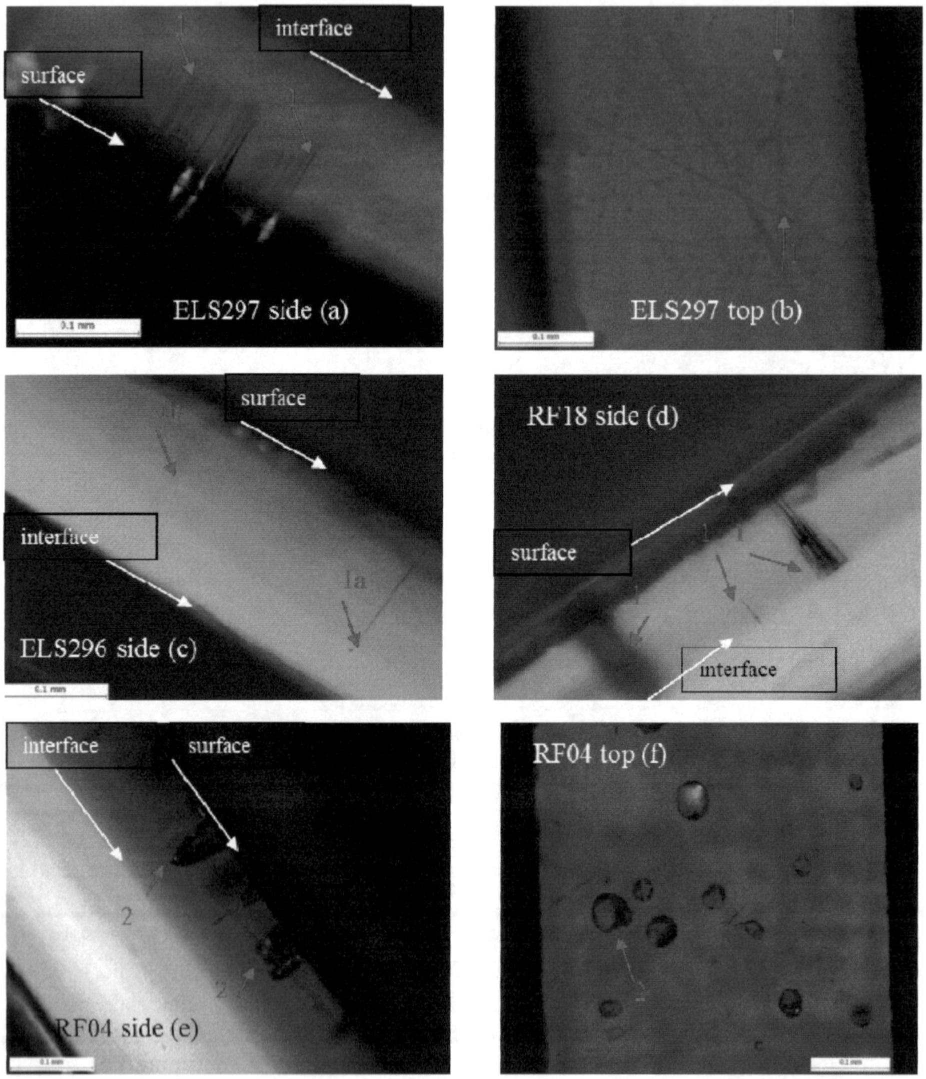

Figure 1. Nomarsky optical images of co-doped 6H-SiC in crossed polarized field (a, c, d, e – side view and b, f – top view); a bar for all pictures is equal to 100 μm. White arrows mark the "interface" between epilayer/substrate and the outer "surface". Red arrows indicate (1) - thin carbon filamentary inclusion, and (2) - polycrystalline cone inclusion. In (a) the colonies of filamentary inclusions are originating from the mid of epilayer thickness. In (c) one filamentary carbon inclusion (1a) is well focused while the other (1b) is out of a focal plane. In (d) the cone shape defects are starting at the interface (the near surface range is unrevealed due to unfinished polishing) and in (e) they appear further within the epilayer. In (f) the polycrystalline inclusions are visible on a top-epi view.

IOP Conf. Series: Materials Science and Engineering **56** (2014) 012004 doi:10.1088/1757-899X/56/1/012004

3. Main Results and Discussion

Figures 2a, 2b show the extracted $\Delta\alpha(t)$ depth-distributions in two samples at different time, *t*, after the laser pulse. At t = 10 ns, an exponential slope from the excited surface originates which can be explained by the absorption generated electron-hole pairs. Band-to-band absorption coefficient value is obtained from the slope (dashed line) is α_{bb} = 900 cm^{-1} which is very close to value of 1000 cm^{-1} reported for undoped 6H-SiC at the same 355 nm excitation wavelength [7]. This shows that band gap absorption is little affected by high co-doping. At the later μs-long time-span, substantial smoothening in the FCA carrier profiles is observed. From the results in figure 1a, b we may conclude that the diffusion driven surface recombination on the excited surface remains small or almost insignificant. This is based on the observation that, at further elapsed times, no carrier gradient arises towards the surface. Low diffusivity is also confirmed by the fact that FCA did not exceed the generation slope on a straight dashed line. These findings are in contrast to substantial ambipolar carrier diffusion phenomenon observed in undoped 6H-SiC [7]. Therefore, the low values of ambipolar diffusion coefficient D = 0-0.1 cm^2/s have been extracted in heavily co-doped SiC by transient grating measurements reported in this issue [see Gulbinas K et al. this issue]. It was shown that, at injection level approaching the concentration of boron, D value start to increase. This most probably promotes easy access for carriers to the recombination centers therefore lifetime decreases.

The injection dependence of the carrier lifetimes in RF18 sample is shown in figure 2c. The FCA amplitude saturation is observed at the 10 μm probe depth at 1 mJ/cm^2 excitation fluence (see inset) which correspond to injected carrier density of about $5 \cdot 10^{18}$ cm^{-3}. The saturation can be attributed to the fact that carrier lifetime becomes shorter than the excitation pulse (τ_R < 2 ns). The FCA smoothing in figure 2a, b is resulting from pronounced injection dependence of the electron lifetime between 2 ns and 1-3 μs.

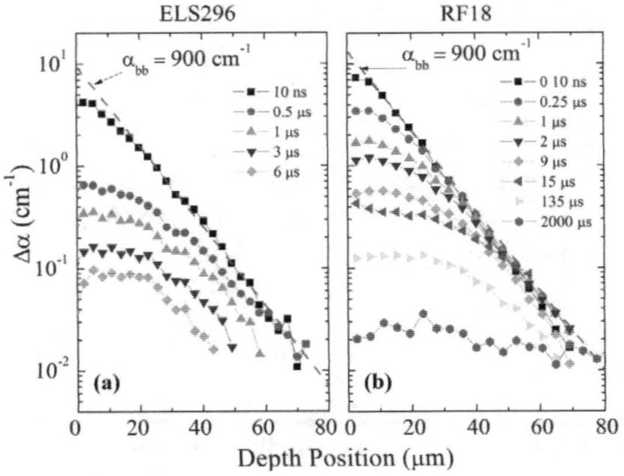

Figure 2a, b Depth-distributions of FCA in two samples at different time after exciting laser pulse.

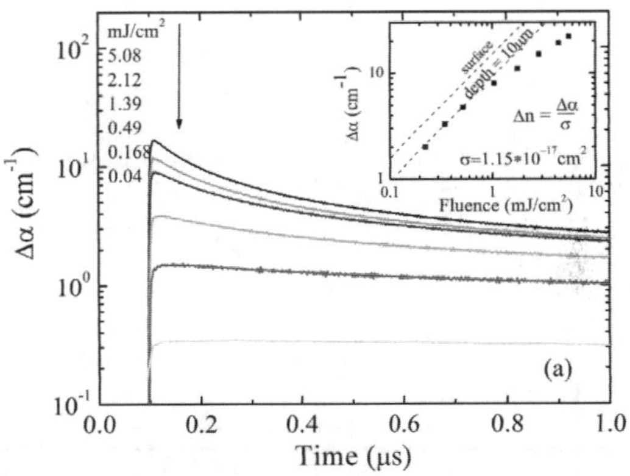

Figure 2c. Induced FCA transients and amplitude (inset) at different excitation fluence in the depth position of 10 μm from the excited surface in RF18 sample. Note that carrier lifetimes are injection dependent.

Figure 3a shows the comparison of $\Delta\alpha(t)$ transients in different samples on a double-log plot at the medium carrier injection of about >10^{17} cm^{-3}. Quite short almost exponential carrier lifetime 0.1-0.18 μs is detected in the high defected samples ELS297 and RF04. While, in two low defective samples (ELS296 and RF18), the carrier lifetime is substantially longer and exhibits components including two extending ones from 50 μs to 1.2 ms. These components are marked by fitted dashed lines in figure 2c. In RF-batch samples the short sub-μs lifetime 0.22 and 1.5 of about same value has been detected by microwave photoconductivity decay at lower injection level of about 10^{16} cm^{-3} (see table 1). This confirms that influence of in-grown defects is related to short carrier lifetime at any external injection provided.

We have performed time-integrated DAP photoluminescence study by excitation with 355 nm 25 ps pulses. At room temperature, PL DAP emission was observed in two samples RF18 and ELS296 containing long carrier lifetimes. However, no PL emission was detected in two samples RF04 and ELS297 with a short lifetime. Since it is known that DAP radiative recombination times are fairly long, appearing on a ms-scale [1], it can be concluded that a short nonradiative carrier lifetime which is related to the grown-in defects prevents DAP emission probability. The investigation of carrier lifetime temperature dependences in a similarly co-doped sample (ELS118: $d_{epi} \approx 45$ μm, N = $9.2{\cdot}10^{18}$ cm^{-3}, B = $5.2{\cdot}10^{18}$ cm^{-3}) has been performed in the orthogonal FCA measurement geometry [Manolis et al. this issue]. It allowed distinction of three recombination paths for free electrons. In figure 3b these relaxation processes are presented schematically. The main non-radiative path is through the $Z_{1/2}$ (carbon vacancy) with microsecond lifetime (figure 3b), thus, it is suppressed due to strongly restricted diffusivity of the free carriers in a heavily co-doped 6H-SiC.

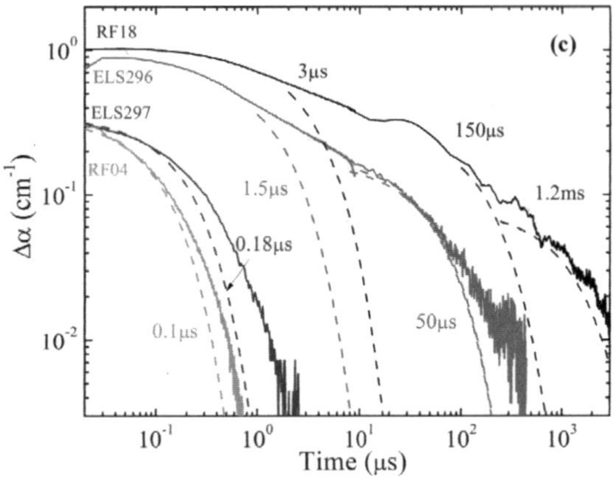

Figure 3a. FCA transients on a double-log scale in four samples at 300 K. Exponential fits are provided by dashed curve with a value of the carrier lifetime.

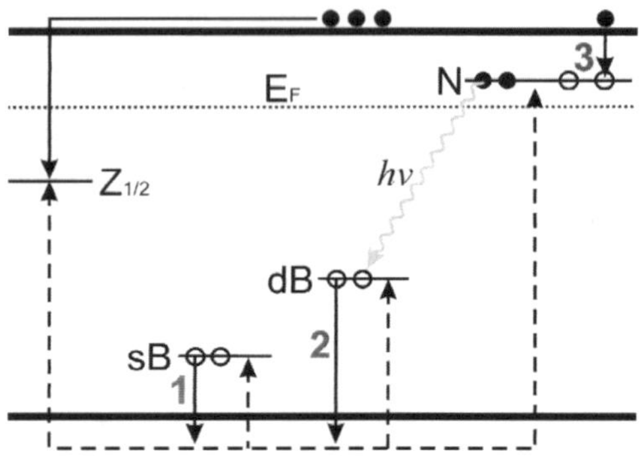

Figure 3b. Schematic of DAP radiation with carrier's release of holes from shallow and deep Boron and to recombination through N sites. Competing recombination channel through $Z_{1/2}$ center is also indicated.

4. Conclusion

The carrier lifetime depth distribution in co-doped B-N 6H-SiC shows that, even at high injection level, the surface recombination driven by carrier diffusion is small and negligible. However due to diffusion increase with injection density the carrier access to the non-radiative centers increases and carrier lifetime obey a drop. This occurs for injection of minority hole density exceeding density of acceptor levels. To achieve efficient DAP luminescence in co-doped samples the ms-long carrier lifetimes are needed. This can be optimized by balancing the doping and improving the crystal quality of the material.

Acknowledgement. This research was supported by Swedish Visby program grant No 00729/2010.

References

[1] Sun J W, Kamiyama S, Jokubavicius V, Peyre H, Yakimova R, Juillaguest S, and Syväjärvi M 2012 *J. Phys. D: Appl. Phys.* **45** 235107
[2] Kamiyama S, Iwaya M, Takeuchi T, Akasaki I, Syväjärvi M, and Yakimova R, 2011 *J. Semicond.* **32** 013004
[3] Kamiyama S, Maeda T, Nakamura Y, Iwada M, Amano H, Akasaki I, Kinoshita H, Furusho T, Yoshimoto M, Kmoto T, Suda J, Henry A, Ivanov I G, Bergman J P, Monemar B, Onuma T, and Chichibu S F 2006 *J. Appl. Phys.* **99** 093108

[4] Syväjärvi M, Müller J, Sun J W, Grivickas V, Ou Y, Jokubavicius V, Hens P, Kaisr M, Ariyawong K, Gulbinas K, Liljedahl R, Linnarsson M K, Kamiyama S, Wellmann P, Spiecker E and Ou H 2012 *Physica Scripta* **T148** 014002
[5] Duijn-Arnold A V, Icoma T, Poluektov O G, Baranov P G, Mokhov E N and Schmidt J 1998, *Phys. Rev.* B **57** 1607-19
[6] Grivickas V and Linnros J 2012 "Carrier lifetime: Free carrier absorption, photoconductivity and photoluminescence" Chapter in the book „*Characterization of Materials*" Second Ed. 3 volums (John Viley & Sons, Inc. US) Ed. Kaufmann E N Vol. 1 pp. 658-692
[7] Galeckas A, Linnros J, Frischholz M, and Grivickas V 2001 *Appl. Phys. Lett.* **79** 365-367

IOP Conf. Series: Materials Science and Engineering **56** (2014) 012005 doi:10.1088/1757-899X/56/1/012005

Raman Scattering and Carrier Diffusion Study in Heavily Co-doped 6H-SiC Layers

K Gulbinas[1], P Ščajev[1], V Bikbajavas[1], V Grivickas[1], O V Korolik[2], A V Mazanik[2], A K Fedotov[2], V Jokubavičius[3], M K Linnarsson[4], M Syväjärvi [3] and S Kamiyama[5]

[1]Institute of Applied Research, Vilnius University, Vilnius, Lithuania
[2]Department of Energy Physics, Belarusian State University, Minsk, Belarus
[3]Dept. of Physics, Chemistry and Biology, Linköping University, Linköping, Sweden
[4]School of Information and Communication Technology, Royal Institute of Technology, Kista, Sweden
[5] Faculty of Science and Technology, Meijo University, Nagoya, Japan

E-mail: Karolis.Gulbinas@ff.stud.vu.lt

Abstract. Thick 6H-SiC epilayers were grown using the fast sublimation method on low-off-axis substrates. They were co-doped with N and B impurities of $\approx 10^{19}$ cm^{-3} and $(4\cdot 10^{16}–5\cdot 10^{18})$ cm^{-3} concentration, respectively. The epilayers exhibited donor–acceptor pair (DAP) photoluminescence. The micro-Raman spectroscopic study exposed a compensated *n*-6H-SiC epilayer of common quality with some 3C-SiC inclusions. The compensation ratio of B through 200 μm thick epilayer varied in 20-30% range. The free carrier diffusivity was studied by transient grating technique at high injection level. The determined ambipolar diffusion coefficient at RT was found to decrease from 1.15 cm^2/s to virtually 0 cm^2/s with boron concentration increasing by two orders.

1. Introduction

It is known that efficiency of the radiative DAP recombination can increase dramatically by introducing large quantities of donors and acceptors in the indirect band gap of 6H-SiC. It is desirable for the co-doped impurities to maintain uniform distribution through the whole thickness of the epilayer, which can reach several hundred microns, and occupy optimal energy states in the 6H-SiC band gap. The emitted DAP light has broad spectrum which, combined with a ultraviolet light excitation from LED of III-V nitrides, can make a preferable white light emitting device with high colour rendering parameters [1]. To solve this problem there is a need to grow SiC on low-off-axis substrates to match GaN and SiC lattice. Unfortunately, under such conditions 3C polytype of SiC also tends to grow. The challenge is to optimise growth conditions in such a way that to avoid 3C-SiC formation and to achieve electronic parameters of 6H-SiC preferable for DAP emission. In this work, some highly co-doped B-N 6H-SiC epilayers were investigated and their characterization was performed in order to study basic requirements for light emitting co-doped 6H-SiC epilayers.

2. Experimental

The epilayers of 6H-SiC were obtained by the fast sublimation growth [2]. The process was performed using (0001) 6H-SiC substrates with the off-orientation angle of 1.4^0 toward $[11\bar{2}0]$ direction. The distance between source and substrate was adjusted to be 1 mm. Polycrystalline SiC plates, doped with different concentrations of B and N, were used as a source material. The nitrogen incorporation was enhanced by growing samples in nitrogen atmosphere. According to secondary ion mass spectrometry profiles (SIMS), the concentrations of N and B in different samples vary in the range of $(9–13)\times10^{18}$ cm^{-3} and $(0.04–8)\times10^{18}$ cm^{-3}, respectively. The epilayers parameters are presented in the Table 1. The backside of each sample was polished to allow ordinary absorption measurements. Some samples were cut into stripes with cross section of $\approx(200\times500)$ μm^2 and length of 5-7 mm. Then both cut edges were carefully polished to allow depth-resolved optical measurements.

Table 1. The parameters of epilayers: doping N, B (SIMS), defects: *a) carbon inclusions; *b) polycrystalline inclusions, ambipolar diffusion coefficient *D* (TG).

Sample No.	Thickness (μm)	N (cm^{-3})	B (cm^{-3})	$N_D - N_A$ (cm^{-3})	Defects (cm^{-2})	D (cm^2/s) at 3×10^{18} cm^{-3}
ELS 117	45	9.0×10^{18}	4.4×10^{18}	4.6×10^{18}	-	0.1
ELS 118	45	9.2×10^{18}	5.2×10^{18}	4.0×10^{18}	-	-
ELS 296	200	1.3×10^{19}	2.1×10^{18}	$\sim1.1\times10^{19}$	10^{1} *a)	0 ± 0.05
ELS 297	200	1.1×10^{19}	4.0×10^{16}	$\sim1.1\times10^{19}$	10^{6} *a)	1.15
RF 18	130	4×10^{18}	2.5×10^{18}	1.5×10^{18}	10^{1}-10^{2} *b)	-
RF 04	110	8×10^{18}	2.2×10^{18}	5.8×10^{18}	10^{5} *b)	-

Unpolarized micro-Raman spectra was collected in the backward direction with the *Nanofinder* confocal system (LOTIS-TII, Belarus - Japan) using excitation of polarized 473 nm solid state cw-laser focused to the spot of ~0.5 μm diameter on the surface. The excitation power was 8 mW and the

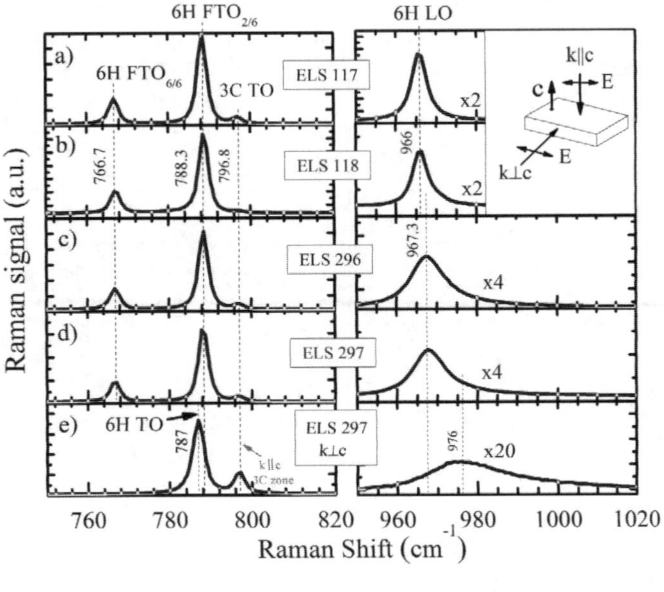

Figure 1. Raman TO-LO lines in co-doped samples at RT. Vertical dashed lines indicate frequencies of TO and LO modes. Dashed red line in (e) demonstrates the result obtained on a large 3C inclusion. The measurement configuration for wave vector of laser light parallel **k**||**c** (a-d) and perpendicular **k**⊥**c** (e) to **c**-axis is shown in the inset.

exposition time 20 s. The ordinary (depth-integrated) absorption was measured using *Shimadzu* UV spectrometer. The depth-resolved absorption was studied by probing across the epilayer (***k***⊥**c**) with a focused 1300 nm cw-laser beam. Carrier lifetime and diffusivity measurements were performed by transient grating (TG) technique at the excitation geometry of ***k***||**c** [3]. The excitation was accomplished by 355 nm and 25 ps pulsed laser. The estimated absorption depth for such wavelength is 11 μm. The period Λ of the TG was varied between 2 and 8 μm. Lifetime τ_R and ambipolar carrier

IOP Conf. Series: Materials Science and Engineering **56** (2014) 012005 doi:10.1088/1757-899X/56/1/012005

diffusivity D were extracted from the plot of inverse TG erasure time τ_g *versus* square of inverse TG period: $1/\tau_g = 1/\tau_R + 4\pi^2 D/\Lambda^2$ [3]. All measurements were performed at RT.

3. Results and Discussion

The selected Raman spectra are given in figure 1 with measurement geometries shown in the inset. All samples exhibit Raman spectra similar in appearance. Everyone consists of one transverse optical TO_0 and two folded TO: $FTO_{2/6}$ and $FTO_{6/6}$ as well as one longitudinal optical LO_0 modes. The position of planar optics Raman modes for the excitation geometry of **k**||**c** and ratio of their magnitude are in close agreement with the data in the available literature [4-6]. In the 3C inclusion (figure 1e, dashed line) the only TO_0 mode at 796 cm^{-1} was observed as was reported in [4,5]. The axial optics LO_0 mode suffers small shifts from 966 cm^{-1} to 967.3 cm^{-1} and a usual broadening with increasing N doping as evident by comparing figures 1a and 1b with 1c and 1d. This broadening, however, is not affected by B doping (note a close similarity between figure 1c and 1d). Larger Raman shift was observed for samples excited from lateral polished side (excitation geometry of **k⊥c,** figure 1e). Moreover, in this case $FTO_{6/6}$ mode disappears due to anisotropy of SiC.

The absorption spectra are shown in Figure 2. All co-doped epilayers exhibit two broad absorption bands at 1.98 eV and 2.8 eV below the band gap (3.0 eV), as in the case of *n*-type 6H-SiC substrate (N of about 10^{18} cm^{-3}). These bands are related to well-known vertical free-carrier-absorption (FCA) transition into two upper conduction band branches. A broad absorption in the infrared region (below 1.0 eV) is caused by phonon-assisted FCA. The value of absorption in different epilayers and substrate tends to be directly proportional to (N_D - N_A) (see Table 1). A small deviation from this tendency at energies between FCA peaks, where absorption is reduced, could be associated with an error in the determination of the epilayer absorption from absorption of epilayer with substrate and absorption of substrate alone. Such behavior of FCA implies that free carrier concentration is determined by (N_D - N_A) and all B atoms serve as compensating impurity. The inset of figure 2 shows depth-resolved absorption at 1300 nm (see A and B points on a vertical line at 0.96 eV), which was recorded by passing light through the lateral polished sides of the sample. Absorption is the highest on the top of the epilayer and gradually decreases towards the substrate by 20-30 %. This trend manifests change of compensation during the process of the epilayer growth.

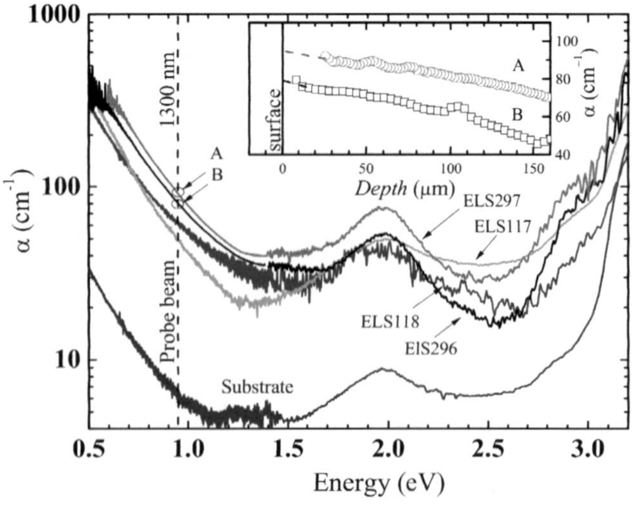

Figure 2. Absorption spectra of different co-doped samples and a substrate. Dashed vertical line marks probe laser energy at 1300 nm. A and B points mark the thickness-averaged absorption in ELS 297 and 296, respectively, while in the inset the absorption depth-distribution coefficient for these samples is shown.

The inset of the figure 3 shows the examples of diffraction efficiency kinetics at different injection fluencies in the sample ELS 296. When excited carrier concentration exceeds the (N_D - N_A) (curve 3 and curve 4), the fast diffraction efficiency component exceeding an order of magnitude has appeared. In the sample ELS 117 this signal is halved and in the sample ELS 297 it is not observed at all. The data from Table 1 led us to propose that the fast signal is largely dependent on the B concentration and could be tentatively attributed to B atom assisted Auger recombination. However, the two-step absorption through the deep B levels could be also considered as a possible cause. Analysis of this fast signal is out of the scope of the present work and will be treated elsewhere.

The recombination lifetime, τ_R, and ambipolar diffusion coefficient, D, obtained by dynamic TG technique for three samples *versus* the injected carrier concentration is presented in Figure 3 (a, b). The values of measured parameters for the injection level of 3×10^{18} cm^{-3} are also included in the Table 1. For all samples both τ_R and D are significantly dependent on the B concentration. For the samples with high B concentration the ambipolar diffusion coefficient is very close to zero (ELS 296, see Table 1 at injection 3×10^{18} cm^{-3}) or is quite small (ELS 117, see figure 3 b). Whereas in the sample with the lowest amount of B (ELS 297) D is substantial. It steeply increases after injection crossing the B concentration level, to value 1.15 cm^2/s at about 3×10^{18} cm^{-3} and further to value 1.6 cm^2/s at the injection of 10^{19} cm^{-3}. This diffusion coefficient increase is not governed by a change in the degeneracy statistics since this effect (in undoped samples 4H-SiC) becomes sufficient at room temperature only at higher injection level above 10^{19} cm^{-3} [10]. We attribute observed trend of diffusion to negatively charged B states that also immobilise holes by trapping them. Consequently, at low injection, both minority holes and majority electrons are strictly hindered by the acceptor-induced space-charge field. Correspondingly, with injection increasing above B concentration, the space-charge field becomes reduced gradually by screening and diffusivity increases to the common values of a free-carrier ambipolar plasma diffusion states [11].

From Fig. 3a is clear the opposite trend occurs with τ_R. In the sample ELS 297 with the lowest amount of B lifetime drops at injection level approaching B concentration. At $\Delta n \geq 10^{17}$cm^{-3} $\tau_R = 0.5$ ns and is independent of injection. For two other samples with high B concentration the carrier lifetime reaches 20 ns and decreases with injection level approaching or above B concentration. It is

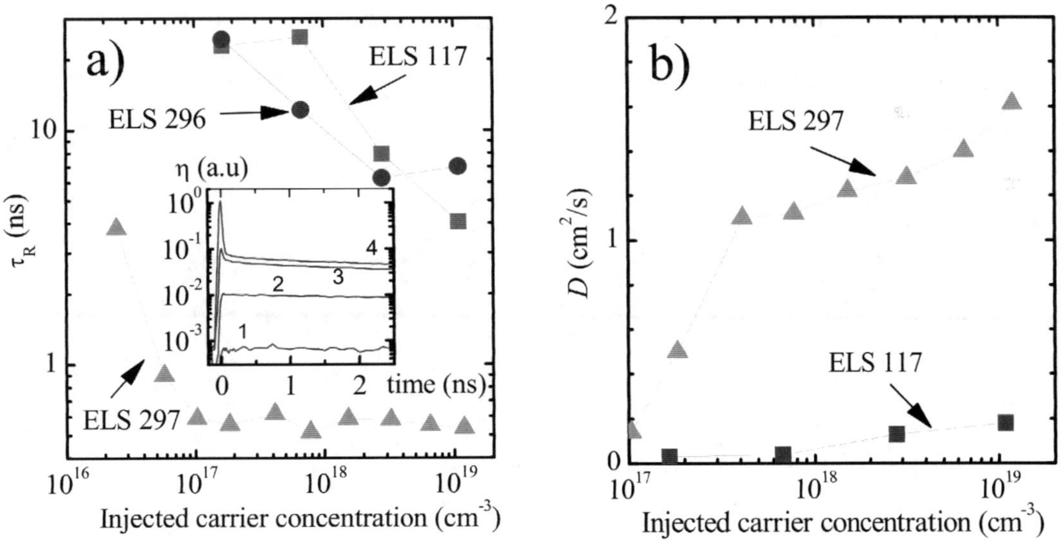

Figure 3. a) Carrier lifetime and b) ambipolar diffusion coefficient *versus* excited carrier concentration in co-doped 6H-SiC samples at RT. The inset shows carrier diffraction efficiency, η, kinetics at increasing excitation fluencies from 0.2 to 12 mJ/cm^2 (1 to 4) in the sample ELS 296.

noteworthy that in the undoped 6H-SiC for similar injection level (3×10^{18} cm^{-3}) the ambipolar diffusion coefficient and lifetime were determined to be $D = 2.7$ cm^2/s and $\tau_R \approx 0.4$ ns [7].

The opposite relationship of diffusivity and lifetime versus injected carrier concentration suggests that carrier lifetime is determined by diffusion-limited recombination times at extended defects. This behaviour is quite common for high band gap compound semiconductors, and is observed in GaN epilayers [8]. It is a consequence of the same effect described above; at low injection and high B concentration diffusivity is restricted due to columbic attraction of minority holes to negatively charged B traps. At high injection holes fill most of B traps and recharge them. Ambipolar diffusivity starts to increase with an increase of the hole and electron concentration providing favourable access of pairs to non-radiative recombination centers. This process compete with more slower channel through N traps which is involved in DAP emission and which is responsible for a hundred μs-long lifetimes at RT [see Manolis et al. in this issue]. This assumption supports the experimental finding that DAP radiative emission at low injection can be steeply enhanced in co-doped 6H-SiC epilayers with N > B> 4×10^{18} cm^{-3} [9].

4. Conclusions

In this study the physical properties of fluorescing 6H-SiC were investigated in order to improve the growth technology of such co-doped layers and to find optimal conditions for the increasing of DAP photoluminescence. The micro-Raman and absorption measurements showed that N-B co-doping does not degrade the quality of crystal structure up to the doping of 10^{19} cm^{-3}. Performing research of the free carriers' diffusivity and their lifetime shows that they both are dependent on B concentration and behave in the opposite way versus injected carrier concentration. Thus, the feasible condition as a contributory factor for DAP emission can be obtained for co-doped 6H-SiC with high B concentration but smaller than that of N concentration. In this case, the restricted diffusivity (most probably due to induced space-charge field) is limiting carrier recombination through nonradiative centres, thus multiple trapping of electrons to N traps stimulates a further radiative emission to deep B levels.

Acknowledgement. This research was supported by Swedish Visby program grant No 00729/2010.

References

[1] Syväjärvi M, Müller J, Sun J W, Grivickas V, Ou Y, Jokubavicius V, Hens P, Kaisr M, Ariyawong K, Gulbinas K, Liljedahl R, Linnarsson M K, Kamiyama S, Wellmann P, Spiecker E and Ou H 2012 *Physica Scripta* **T148** 014002
[2] Syväjarvi M and Yakimova R, in SEST ed. by Bhattacharya P, Fornari R and Kamimura H 2011 ISSN 978-0-444 53144
[3] Ščajev P, Jarašiūnas K, Kadys A, Storastas J, Abramov P L, Lebedev S P, and Lebedev A A, 2010 *AIP Conf. Proc.* **1292** 107
[4] Pusche R, Hundhausen M, Ley L, Semmelroth K, Schmid F, Pensl G, and Nagasawa H 2004 *J. Appl. Phys.* **96** 5569
[5] Nakashima S and Harima H 1997 *Phys. Stat. Sol. (A)* **162** 39
[6] Nakashima S and Tahara K 1989 *Phys. Rev. B* **40** 6339
[7] Tamulaitis G, Yilmaz I, Shur M S, Anderson T, and Gaska R 2004 *Appl. Phys. Lett* **84** 355
[8] Ščajev P, Usikov A, Soukhoveev V, Aleksiejūnas R, and Jarašiūnas K 2011 *Appl. Phys. Lett.* **98** 202105
[9] Sun J W, Jokubavicius V, Liljedahl R, Yakimova R, Juillaguet S, Camassel J, Kamiyama S, Syväjärvi M 2012 *Thin Solid Films* **522** 33
[10] Ščajev P and Jarašiūnas K, 2013 *J. Phys. D.: Appl. Phys.* **46** 265304
[11] Grivickas P, Martinez A, Mikulskas I, Grivickas V, Tomašiūnas R, Linnros L and Lindefelt U 2001 *Mater. Sci. Forum* **353-356** 353

IOP Conf. Series: Materials Science and Engineering **56** (2014) 012006 doi:10.1088/1757-899X/56/1/012006

Temperature Dependencies of Free-Carrier-Absorption Lifetime in Fluorescent 6H-SiC Layers

G Manolis[1], K Gulbinas[1], V Grivickas[1], V Jokubavičius[2], M K Linnarsson[3], and M Syväjärvi[2]
[1]Institute of Applied Research, Vilnius University, Saulėtekio av. 10, Vilnius, LT-10223 Lithuania,
[2]Department of Physics, Chemistry and Biology, Linköping University, Linköping, SE-58183 Sweden
[3]School of Information and Communication Technology, Royal Institute of Technology, SE-16440 Kista, Sweden

E-mail: g.k.manolis@gmail.com

Abstract. The nonradiative decay of majority electrons has been studied over a wide temperature range from 80 K to 600 K using the time-resolved free-carrier-absorption (FCA) technique. At high injection level of the highly-luminescent N-B codoped 6H-SiC epilayer, we revealed three main relaxation components of injected free electrons over ps-to-ms time ranges. By means of temperature dependency, two components can be ascribed to thermal activation of holes from a shallow (200 meV) and a deep (500 meV) acceptor. The third one, which has a hundred μs-time scale, we attribute to minority hole recombination from the valance band into the electron trap (53 meV). This recombination channel seems to compete with the deep-acceptor (Boron) to-donor (Nitrogen) pair visible emission at and below 300 K.

1. Introduction

The need for artificial daylight illumination has driven research to the concept for white light-emitting diodes (LEDs). The spread LED models are based on III-V nitrides with a yellow phosphor coating acting as a light converter. These phosphors require rare-earth metals which are becoming scarce. The problem becomes pronounced because of a high color temperature and too low color-rendering index [1]. The 6H-SiC can be a substrate for growth of GaN LED structures to growth. On the other hand, if properly doped, such SiC can be composed of two different co-doped layers (B-N and Al-N) acting as light converters that emit into two visible broad spectral light ranges by an efficient donor-to-acceptor pair (DAP) recombination. Such 6H-SiC B-N epilayer have already have been demonstrated and thus show a new potential especially when the material quality is improved significantly [2]. Therefore, among remaining scientific issues, there is a fundamental question if the competing nonradiative process in heavily co-doped SiC material can be minimized so that it will not interplay greatly with the DAP recombination channel.

In this work epitaxial co-doped B-N 6H-SiC layers have been grown by the fast sublimation growth process and studied by time-resolved FCA technique under high injection in order to reveal the relation of non-radiative recombination processes with existing trap levels and their relation to the main dopants of the material.

2. Samples and experimental techniques

The fast sublimation growth process was used to grow homoepitaxial layers on low-off axis 6H-SiC substrates. Nitrogen (N) and boron (B) were introduced into epilayer via controlled co-doping from the polycrystalline SiC source and the N concentration was also enhanced by the gas ambient during the growth. The epilayer thickness was $d \approx 45$ μm and the concentration of N and B dopants was established by secondary ion mass spectrometry in the sample (ELS118) to be 9.2×10^{18} and 5.2×10^{18} cm^{-3}, respectively [1].

The FCA measurements have been performed from 80 K to 600 K in an open cycle liquid nitrogen cryostat with a heater. The sample was excited from the epi-surface side with 8 ps duration pulses of the third harmonic from a Nd^{+3}:YLF laser (351 nm) operating at 10 Hz. The pump light creates a dense electron-hole plasma with exponential in-depth distribution amid a penetration depth 1/e = 11 μm which is nearly independent of temperature. The carrier density in the substrate can be virtually neglected. The FCA probing was performed in a collinear geometry with delayed pulses at either 1053 nm or 1064 nm wavelength in the transparency region of 6H-SiC. The delay of the probe was implemented in two ways: optically, by changing the optical path of the ps-probe (to 4 ns), or electrically, when another YAG laser with 4 ns pulse was electrically synchronized with the pump laser (delay range ≈ 200 μs).

3. Experimental results and analysis

To show the complete time range, in Fig 1a we plot the FCA signals [$\ln(T_0/T) = \Delta\alpha d$] on a log delay-scale. The FCA maxima are artificially set at 10 ps. These data are measured at indicated temperatures for a constant 40 mJ/cm^2 pump flux that results in a 7×10^{19} cm^{-3} concentration of the excess electron-hole (e-h) pairs over the absorption depth, e.g. above the doping levels for about the half of the epilayer thickness. The "fast" decay (for $t < 0.04$ ns) is observed at this excitation level but does not appear for excitation below the doping level (not shown). Its amplitude and relaxation time is T-independent. The appearance of the "fast" part is not well understood; therefore we tentatively attribute it to the impurity-assisted Auger processes. In this study, we will focus on the extended "slow" decay parts that, in n-type 6H-SiC, most probably are related to the recombination processes of remaining majority excess electrons in the conduction band. From Fig. 1a it can be depicted that components in the "slow" decay are strongly T-dependent. Multiple three decay processes might be presented after analysis of decays curves. This is demonstrated by exponential fits (dashed lines) in Fig. 1b at 200 K and similarly at other temperatures. The fitting is satisfactory taking into account that the trap occupation is not complete by excitation at the deep depth of the epilayer.

The fitting parameters, namely the relaxation time and the component amplitude are presented in Fig. 2a,b into an Arrhenius diagram. It should be stated that all relaxation times are separated by an order of magnitude in value, or more. This allows the assumption that the obtained components do not

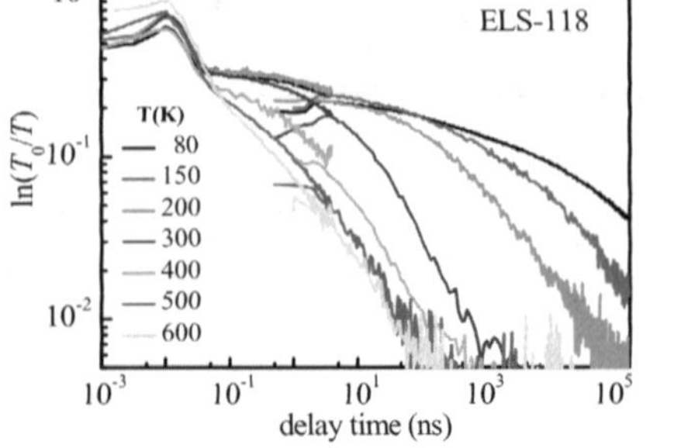

Figure 1a. FCA signal as a function of log delay time at different T. An overlapping probing for optical/electrical is at $t \cong 4$ ns. T-dependent "slow" part can be viewed.

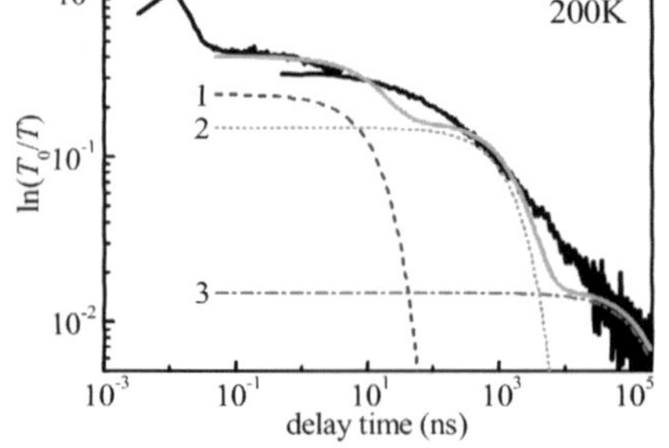

Figure 1b. Fitting of three independent exponents to the experimental data at 200 K: Labels "1", "2" and "3" according to the description in the text.

affect each other and, thus, it is reasonable to apply a simple exponential fitting, as in Fig. 1b. From results in Fig. 2 we see that two faster components denoted as "1" and "2" have a strong T-deactivated relaxation time but weakly T-dependent amplitudes. Whereas the component "3" shows clearly an opposite case; T-independent relaxation time and strongly thermally deactivated amplitude.

Since the negatively charged acceptor is a minority carrier trap with much smaller capture cross section for majority electrons, consequently, the hole trap occupation in the forbidden energy band gap is of the major importance between 400-600 K. It is known that B atoms are acceptors in SiCs and can have positions with different ionization levels [3]. When we depict a linear fit by solid lines to the obtain results in Fig. 2a, these extracting values give thermal deactivation energies $E_{a1} \approx 200\pm30$ meV and $E_{a2} \approx 500\pm120$ meV. The large scatter for E_{a2} is caused by the "fast" component at high temperatures which actually was subtracted artificially from the plot of Fig. 2a. The line fitting in this T range, however, is of sensitive to the subtracted amount.

Since B is the dominant acceptor impurity in the investigated epilayer [1], both the activation energies should be compared to B acceptors, one for a shallow level (sB) and one for a deep level (dB). The obtained experimentally values are lower than the actual energies reported for sB (220-350 meV) and for dB (550-750 meV) [4]. It should be noted that the dB level is related to a complex center on carbon sublattice but distributed over a smaller volume than in the case of the sB acceptor. It can also possibly be located either at an *h*-site or on a *k*-sites [3]. The discrepancy between our findings and reported values can arise due to three reasons. The first one is the B (as well as N) concentration probably induces a band-gap shrinkage and, partly, the particle level spread into an impurity energy band. The second reason may arise from the fact that close distance dB-N pairs can transform the dB complex structure. The third one is that we have omitted the re-capture process for holes in our consideration. As the temperature increases, the capture results in an apparent activation energy that is lower than the trap level [5]. However, the above determined energy values for minority carriers can be a proof that B impurity atoms are perhaps responsible for those decay components. Moreover, as in Fig. 2b the signal amplitudes from each independent mechanism, that correspond to B levels, reveals that the concentration of shallow centers is slightly exceeding that of deep ones as it is reported for epitaxially grown layers [4].

The third relaxation mechanism, denoted as "3" in Fig 2, has shown a potential effect at 300 K and below. Observed carrier trapping that features the amplitude reduction without changing relaxation time has been analysed by Ichimura [5]. According to provided his analysis, the situation can be attributed to minority carrier recombination through the majority carrier trap in the upper part of the forbidden band gap. Once electrons are emitted from the trap, the trap cannot capture minority holes.

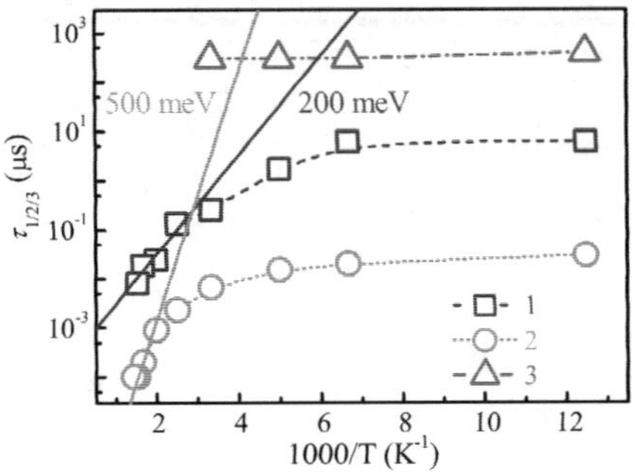

Figure 2a. Arrhenius plot for the three independent relaxation processes of the "slow" decay. Extracted thermal activation energy of component "1" and "2" is indicated by a solid line fit at high T.

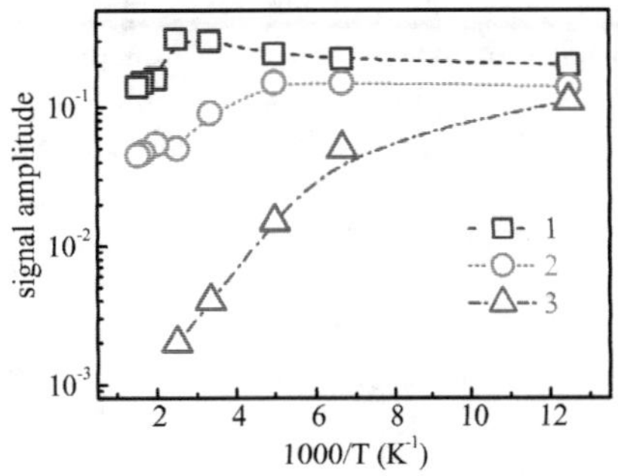

Figure 2b. Signal amplitudes of three independent processes. While component "1" and "2" amplitudes are weakly T-dependent, the component "3" amplitude is strongly T-dependent.

Then the occupation probability $f = 1/[1+g\cdot\exp(E_a - E_F)]$ is given for thermal equilibrium, where g is the degeneracy of unpaired electrons. Noting $n = N_C\cdot\exp\{(E_F - E_C)/kT\}$ is a constant, one can arrive to $(1 - f)/f = (N_C\cdot g/n)\cdot\exp\{(E_a - E_C)/kT\}$. A plot in Fig. 3a shows the amount of the "3" component amplitude normalized by that at the low temperature limit denoted as R (where the factor $T^{-3/2}$ corrects for the temperature dependence of N_C). This plot reveals that thermal activation energy of this mechanism is $E_{a3} \approx 53\pm5$ meV. This value is true for N atoms close to conduction band in heavily doped 6H-SiC but lower than the atomic level at h-site for low doping (80 meV [4]). We note that a carrier recombination process on N site is indirect, and hence, for a detailed analysis one has to take under account phonons contributions which we have omitted in the present study. In Fig. 3b the three relaxation processes which have been discussed at previous paragraphs are presented schematically.

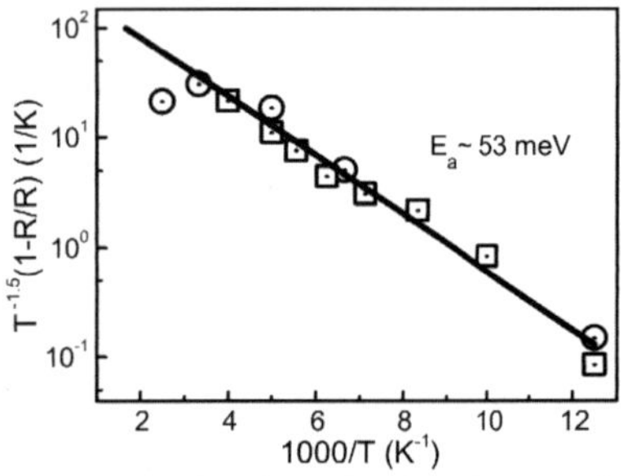

Figure 3a. Arrhenius plot for the amplitude of the "3" component (Fig. 2b) where R is the amount of amplitude normalized by that at the low temperature limit.

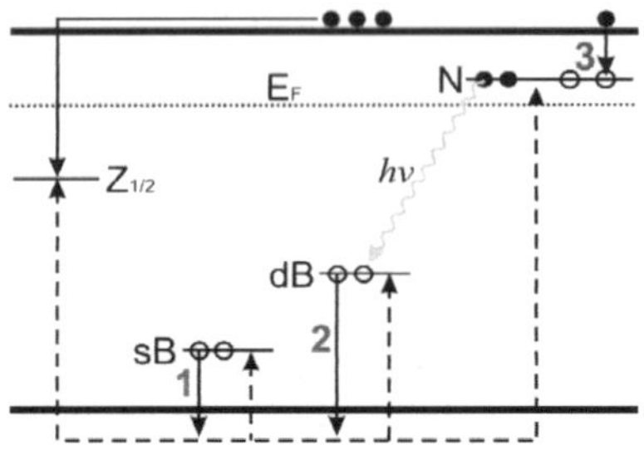

Figure 3b. Schematic of DAP radiation with carrier's relaxation processes "1", "2" and "3" denoted to thermally emitted holes from shallow and deep Boron and to recombination through N sites. Competing recombination channel through $Z_{1/2}$ center is also indicated.

4. Conclusions

By applying time-resolved FCA technique we have shown that long living decay processes in B-N co-doped 6H-SiC are composed of three main components, which are well separated on a time-scale and obey quite different T-dependences. The "1" and "2" component can be ascribed to accelerated nonradiative recombination rate due to hole thermal emission from shallow and deep levels of B with thermal activation energies determined $E_{a1} \approx 200$ meV and $E_{a2} \approx 500$ meV. The third component, which is responsible for the far longer decay times, can be ascribed to free holes recombination with electrons at the indirect trap related to N with a thermal energy of $E_a \approx 53$ meV. This lifetime is competing with a radiative DAP emission rate. Thus, all observed recombination channels are related to the main dopants that are introduced. We assume that there also exists a competing recombination through the known $Z_{1/2}$ centers with μs-time scale (Fig. 3b), therefore, it is suppressed due to strongly restricted diffusivity of the free carriers in a heavily co-doped 6H-SiC [see Gulbinas, this issue].

References

[1] Syväjärvi M, Müller J, Sun J W, Grivickas V, Ou Y, Jokubavicius V, Hens P, Kaisr M, Ariyawong K, Gulbinas K, Liljedahl R, Linnarsson M K, Kamiyama S, Wellmann P, Spiecker E and Ou H 2012 *Physica Scripta* **T148** 014002

[2] Syväjarvi M and Yakimova R, in SEST ed. by Bhattacharya P, Fornari R and Kamimura H 2011 ISSN 978-0-444 53144

[3] Duijn-Arnold A V, Icoma T, Poluektov O G, Baranov P G, Mokhov E N and Schmidt J 1998, *Phys. Rev. B* **57** 1607-19

[4] Lebedev A 1999 *Semiconductors* **33** 107-30

[5] Ichimura M 2006 *Solid-State Electronics* **50** 1761-6

Passivation of SiC device surfaces by aluminum oxide

A. Hallén[1]*, M. Usman[2], S. Suvanam[1], C. Henkel[1], D. Martin[1] and M.K. Linnarsson[1]

[1]Royal Institute of Technology KTH-ICT, P.O. Box Electrum 229, SE 164 40 Kista, Sweden

[2]Experimental Physics Laboratories, National Centre for Physics, Islamabad, Pakistan

*E-mail: ahallen@kth.se

Abstract. A steady improvement in material quality and process technology has made electronic silicon carbide devices commercially available. Both rectifying and switched devices can today be purchased from several vendors. This successful SiC development over the last 25 years can also be utilized for other types of devices, such as light emitting and photovoltaic devices, however, there are still critical problems related to material properties and reliability that need to be addressed. This contribution will focus on surface passivation of SiC devices. This issue is of utmost importance for further development of SiC MOSFETs, which so far has been limited by reliability and low charge carrier surface mobilities. Also bipolar devices, such as BJTs, LEDs, or PV devices will benefit from more efficient and reliable surface passivation techniques in order to maintain long charge carrier lifetimes. Silicon carbide material enables the devices to operate at higher electric fields, higher temperatures and in more radiation dense applications than silicon devices. To be able to utilize the full potential of the SiC material, it is therefore necessary to develop passivation layers that can sustain these more demanding operation conditions. In this presentation it will also be shown that passivation layers of Al_2O_3 deposited by atomic layer deposition have shown superior radiation hardness properties compared to traditional SiO_2-based passivation layers.

1. Introduction

Reliability and long-term stability are coming more into focus, as silicon carbide devices are being introduced on the market. Typical reliability issues concern passivation and packaging of semiconductor device structures and in this contribution we will discuss problems related to the passivation layers and the processing of such layers. Important physical properties of a passivating layer on a semiconductor are large dielectric constant, large bandgap and sufficient band offsets between the valence and conduction band edges between the passivating dielectric layer and the semiconductor. A high thermal conductivity and a thermal expansion coefficient of similar magnitude as that of the semiconducctor are also desirable. Apart from these properties of the layer itself, it is critical to select a deposition technique and surface pre-treatment that can enhance the overall reliability and electric field durability of the system. Very important for surface passivation is that the interface and near interface regions between the dielectric and the semiconductor contains a low and thermally stable concentration of electronic defects. The presence of such surface defects alters the surface potential, decrease the mobility and increase the recombination of charge carriers, leading to an overall degradation of device performance, both for uni- and bipolar electronic devices, and also optoelectronic devices. Since ideal, defect free, interfaces does not exist in reality, the technology must aim at minimizing the amount of defects and, maybe even more important, ensure that the defects does not change over time due to thermal stress, electric fields, radiation, or other ambient conditions. Finally, when optimizing passivation technology, it is also important to assess the process integration and compatibility with other SiC processing steps.

The passivation layers used today for commercial SiC devices are mostly inherited from silicon technology, i.e., SiO_2. Silicon dioxide is still adequate for the thermal specification and voltage range of todays commercial SiC unipolar devices, although the breakthrough for SiC MOSFETs is severely hampered by the SiO_2/4H-SiC gate properties. For bipolar devices the surface recombination is a major issue, for instance in the region between the mesa-etched n-type emitter and the p-type base contact in 4H-SiC bipolar transistors. In this case, the increased surface recombination of a poorly passivated surface strongly lowers the transistor gain [1,2]. Also opto-electronic devices, such as LEDs, or future photovoltaic devices, will suffer from surface recombination, resulting in lower efficiency.

The basic physical behaviour of the SiO_2/SiC interface has been extensively studied with pioneering work, for instance by Afanas'ev [3], and there exist today a relatively well established physical model for the interface and near interface regions between SiO_2 and the three most common SiC polytypes, 3C, 4H, and 6H. In this model there are two types of defects related to the excess of carbon atoms at the interface, as the Si atoms are consumed in the oxidation process. These defects will give rise to bandgap states which will influence charge carrier transport at the interface. The first type are π-bonded carbon clusters of varying sizes yielding states close to the conduction band and also states distributed through the entire bandgap. Smaller clusters will have levels located close to the conduction band edge, while larger, graphitic-like clusters are more evenly distributed in the bandgap. The smaller C-clusters are also believed to be the reason why SiO_2/4H-SiC interfaces presents more difficulties than interfaces with 3C- or 6H-SiC polytypes; since the 4H polytype has a larger bandgap, more of the small C-clusters will be present within the bandgap in this polytype. In 3C and 6H, the states appear in the conduction band and will not be as influential for the charge carrier dynamics. The second type of defects appear on the oxide side of the interface and is often called near interace trap, NIT. The chemical origin is still under discussion, but oxygen vacancies and other intrinsic defects in the oxide have been proposed [4-7].

Large engineering efforts have been made to improve the SiC/SiO_2-interface and considerable progress has been made [8]. The techniques that have been attempted have again, to a large extent, been inherited from Si-technology. Many techniques to prepare the surface prior to oxidation has been reported, for instance by UV-exposure [9], ozone treatment [10] and by combing ozone and UV-cleaning [11], but also implantation of nitrogen and other ions to tailor the chemical content at the surface [12-15]. Passivation of dangling bonds at the interface have, for instance, been tried with hydrogen [16] and hydrogen peroxide annealing [17] with some progress, although it seems difficult to assure a thermal stability for these techniques. So far the most fruitful method, and also the most commonly used technique today, is post-oxidation annealing in NO or N_2O [18-21]. Although the N-treated SiO_2/4H-SiC interfaces have greatly improved the interface quality, the properties are still not as good as for the SiO_2/Si system. For instance, the electron surface mobility in the latter case is about 40% of the bulk mobility, while the electron surface mobility for the SiO_2/4H-SiC interface is around 10-15% of 4H-SiC bulk mobility. For the density of interface, or near interface, traps concentrations for the SiO_2/Si interface is in the low 10^{10} $cm^{-2}eV^{-1}$ and at least one order of magnitude higher for SiO_2/4H-SiC. Regarding the long term and high temperature reliability, the N-annealed SiO_2/4H-SiC systems also needs improvements, especially when going to higher temperature operation where the full benfits of SiC can be utilized.

Although there may be further room for improvements of passivation of 4H-SiC by SiO_2, it is not certain that this dielectric is the best choice for SiC, for instance due to the relatively low dielectric constant of 3.8. Several studies show good results with other dielectrics of higher k-values, such as AlN [22,23], Al_2O_3 [24-28], HfO_2 [29,30], but for the reliability it is also important that the band offset is large enough to reduce tunneling across the barrier. The problem is that the dielectric constant is generally inversly proportional to the bandgap, which means that alternative dielectrics with higher k-values will result in smaller offset and increased leakage. This problem could be overcome by combining layers in dielectric stacks [26,28,29,31], utilizing individual properties of each layer to optimize the overall stack performance. A highly interesting alternative dielectric is aluminum oxide, Al_2O_3, deposited by atomic layer deposition (ALD). Aluminum oxide has a similar bandgap as SiO_2, but has about doubled dielectric strength. It is also very well suited for the ALD process which has several advantages compared to other deposition processes. There are indications that by using this dielectric on 4H-SiC it is possible to avoid the traps in the vicinity of the conduction band (NITs)[27] and results in our group with Al_2O_3 has confirmed the promising properties and show that the electrical performance can be at least as good as that of SiO_2. In addition, it has recently been shown that Al_2O_3 has a much better ability to withstand ionizing radiation than SiO_2 [32,33], and this can be very important since one of the potential applications of SiC devices is in radiation rich environments.

In this report we will have a closer look at Al_2O_3 on n-type 4H-SiC, which will be compared to SiO_2. In Table I the properties two Al-based dielectric materials are compared to SiO_2. The table also shows values for AlN, although this material is not discussed further. The focus will be more on engineering and processing aspects -how- than on physical modelling and detailed understanding - why. It will, for instance, be shown that although "standard cleanroom procedures" are performed to clean the surfaces prior to deposition of the dielectric layer, very different results are obtained for slightly different cleaning procedures.

Table 1. Comparison of some physical properties for SiO_2, Al_2O_3 and AlN, static relative dielectric constant, ε_r, band gap, E_g, difference between dielectric conduction band edge and Si and 4H-SiC conduction band edges, and thermal conductivity, κ.

Dielectric	Static relative dielectric constant ε_r	Band gap E_g (eV)	Conduction band offset from Si/4H-SiC (eV)	κ (W/cmK)
SiO_2	3.9	9	3.5/2.7	0.015
Al_2O_3	8	8.8	2.8/1.7	0.02
AlN	9	6	2.2/1.7	19

2. Experimental methods

For the experiments, n-type epitaxial 4H-SiC grown on 4° off axis substrate wafers from SiC Crystal AG are used. Typically, the nitrogen doping of the epitaxial layer is in the range of 10^{15} to 10^{16} cm^{-3} and the layer thickness is about 5-10 μm. One of the objectives of this study is to optimize the surface pre-treatment prior to depostion of the dielectric. The investigated predeposition cleaning procedures include three types, here named A, B, and C. The first surface cleaning procedure (A) is a standard 7 Up/IMEC clean consisting of 5 minutes in 3 l HSO_4: 1 l H_2O_2 solution followed by a de-ionized (DI) water rinse, then 100 second IMEC clean in 6 l DI H_2O: 60 ml isopropanol : 60 ml HF (50%) solution and, lastly, a rinse again in DI water. Procedure B consist of a 10 minute week RCA1 clean in 100 ml NH_4 : 2 l DI H_2O : 400 ml H_2O_2 solution at 60 °C followed by a rinse in DI water then a dip for 30 seconds in HF (1:50) and, lastly, rinsed in DI water for 60 seconds. Finally, procedure C consists of annealing in an Ar ambient at 400 °C for 1 hour in the ALD deposition chamber prior to aluminiun oxide deposition. In cases B and C the original samples have undergone procedure A prior to the additional surface processing steps. It is expected that procedure A should clean the surface of all contaminating metals and organic material and removal of the resulting oxide layer, procedure B further removes organic contaminantes and any resulting oxide layer. Finally, the annealing procedure included in C may further removes contaminantes from the sample surface. It is hoped that the differences in surface cleaning procedure will result in a beneficial passivation technique for the SiC surface. For example the organic clean in procedure B could help in reducing the negative electrical charge effects of carbon clusters in the dielectric/4H-SiC interface.

Aluminum oxide is deposited through the atomic layer deposition process (ALD) using trimethyl aluminum (TMA) and H_2O at 200 °C for 450 cycles for an aluminum oxide thickness of about 50 nm. The films are deposited using a BENEQ TFS 200 ALD deposition system. The films are subsequently annealed at 500 °C for 60 seconds in N_2O by rapid thermal annealing. Reference samples with silicon oxide are formed by depositing the layer using PECVD at 300 °C, followed by N_2O annealing at 1150 °C for 60 minutes. Reference samples with thermal oxide, grown at 1250 °C in N_2O, are also used. Back side contacts are provided by Ni silicide. The wafers are cut in 1 cm^2 samples and top contacts of varying sizes are formed by Al deposition to complete the MIS structures.

The characterization of the samples is mainly performed by electrical measurements, but structural properties and composition of the layers is also measured by scanning electron microscopy, elipsometry, X-ray diffraction, and Rutherford backscattering spectrometry. These results have been published elsewhere [34]. Electrical characterization is performed in darkness and under controlled temperatures by current voltage (IV) and capacitance voltage (CV) measurements. A probe station together with a HP4156 parameter analyser and a HP4284A LCR meter is used for the IV and CV measurements, respectively.

3. Results and discussion

In order to optimize the cleaning procedures before deposition of dielectrics, three different commonly used cleaning recepies are compared. The details for these procedures, A, B, and C, are described in the previous section and the resulting CV, measured at room temperature, is shown in Fig. 1 for the three different cases. In Fig. 2 we show the values extracted from the CV data in Fig. 1 for dielectric constant, ε_r, flatband voltage shift, V_{fb}, and density of interface states, D_{it}, the latter estimated by the Terman method. It is obvious that the three different pre-deposition treatments have profoundly different influence on the capacitance of the structure. Procedure B provides the highest dielectric constant (10.2) lowest flatband voltage shift (3.4 V) and lowest D_{it} (1.6×10^{11} cm^{-2}), while procedure C displays the worst values for these parameters, ε_r=7.4, V_{fb}=5.9 V, and D_{it}=2.9×10^{11} cm^{-2}.

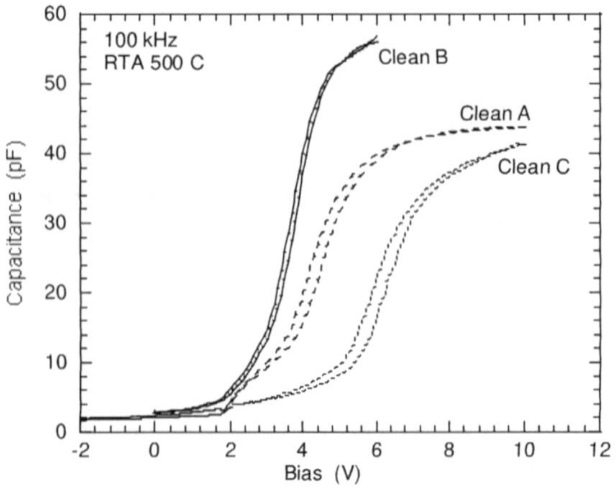

Figure 1. Capacitance versus voltage measurements of MIS structures with ALD-Al_2O_3 for three different types of surface pre-deposition cleaning procedures, A, B and C, further described in the text.

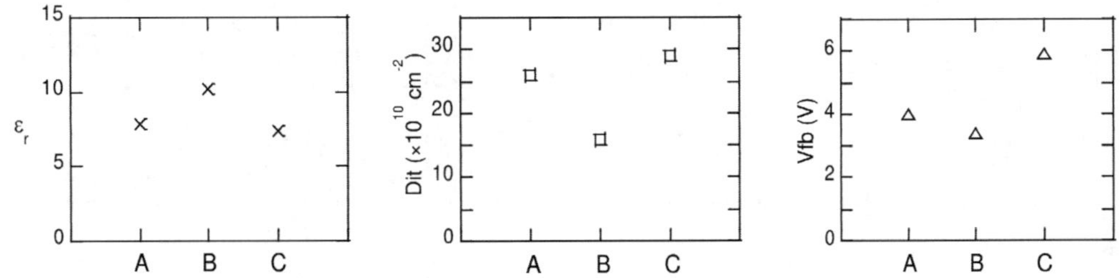

Figure 2. Values extracted from Figure 1 for a) relative dielectric constant, b) density of interface trap and c) flatband voltage shift for the three different cleaning procedures A, B, and C.

It appears that adding the weak RCA clean to the 7Up/IMEC procedures, provides the most efficient removal of contaminants and natural SiO_2, which ensures a homogeneous Al_2O_3 to form. Cleaning procedure C, including "baking" of the sample in the deposition chamber before deposition, failed, possibly due to insufficient vacuum in the chamber, about 10^{-6} torr. This vacuum may result in a net deposition of contaminants instead of removal. The results obtained also highlights the importance of controlling the basic processes and handling of samples in clean room environment. Such elementary procedures as wafer cleaning often have a stronger influence on the final results than introduction of new dielectrics and ingenious combinations of layers.

The choice of dielectric is of course based on the physical properties of the layers, but the choice is also intimately coupled to the deposition method. Atomic layer deposition is a very attractive method for deposition of thin films with excellent control of thickness over large areas at relatively low temperatures. Deposition of Al_2O_3 by ALD is in some sense a model example for the technique and the combination of the excellent physical properties of the layer and the deposition technique is a strong motivation for using this system. One problem for Al_2O_3 is the fixed charge, mostly negative [27], which seems to be related to intrinsic oxide defects. This charge shifts the flatband voltage, V_{fb}, in the CV curve to larger positive values. Figure 3 shows an extreme case of flatband voltage shift of about 20 V found in an early sample. It is also seen in the figure that the negative charge can be removed by a thermal anneal in 950 °C in Ar atmosphere for 1 hour. The cause for the negative charge is not known, but it is most likely related to the pre-deposistion treatments and inadequate removal of native SiO_2. However, such a long and high temperature annealing will affect the structure of the Al_2O_3, which will become more crystalline. This results in grain boundaries and increased current leakage.

For SiO_2 on n-type 4H-SiC the oxide charge is most often less than for as-deposited Al_2O_3. The sign of the charge in the SiO_2 case also varies and possible cause for this is that two dominating types of defects, one with positive charge and one with negative charge, compete. Different pre-deposition treatments, as well as growth techniques will decide which defect that will dominate [18]. Figure 4 is an illustration of this, showing CV measurements of a MIS structure consisting of 4H-SiC n-type epi with oxide either deposited by PECVD at room temperature followed by a post-oxide anneal in N_2O at 1150 °C, or grown in N_2O at 1250 °C. The samples have identical cleaning procedures prior to the oxidation process. The PECVD deposition process results in positive charges possibly due to charges fixed in the oxide, while the N_2O-grown oxide shows negative charge, which could be due to acceptor like traps close to the interface..

IOP Conf. Series: Materials Science and Engineering 56 (2014) 012007 doi:10.1088/1757-899X/56/1/012007

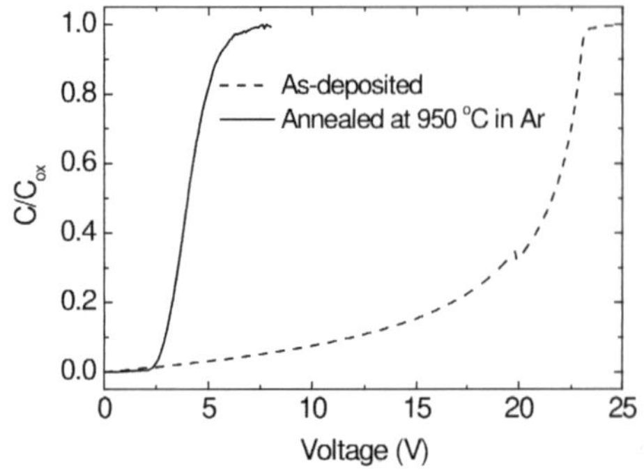

Figure 3. CV measurements of a MIS structure with ALD-Al_2O_3 before and after annealing in Ar atmosphere at 950 ºC for 1 hour. The fixed negative charges in the oxide, or at the interface, gives a flatband voltage shift of 21 V, but following an anneal, most of this charge is neutralized.

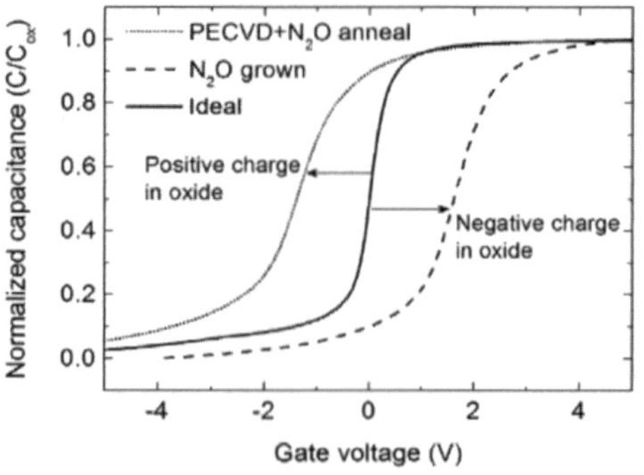

Figure 4. CV measurements of a MIS structure consisting of n-type 4H-SiC epitaxial layers with SiO_2 deposited by PECVD and thermally grown in N_2O at 1250 ºC. The PECVD deposited oxide is annealed in N_2O at 1150 ºC. An ideal CV curve for the structure without flatband voltage shift is also shown.

Finally, in this brief review, we will consider the electrical breakdown properties of the dielectric. Such measurements are time consuming and often destructive and also needs to be performed on a large number of samples, since the breakdown behavior can differ substantially due to local weak spots from defects or inhomogeneous processing. In order to understand the mechanisms involved in the current conduction, it is also needed to perform measurements at different temperatures. Here we present a comparison between the typical breakdown behaviors for SiO_2 and Al_2O_3 dielectrics on n-type 4H-SiC epitaxial structures. The silicon dioxide samples are the same PECVD samples from which the CV characteristics are shown in Fig. 4, while the Al_2O_3 MIS structures are formed by ALD as described in the experimental section. The samples undergo identical cleaning prior to the dielectric formation using procedure A. The leakage current is shown as a function of forward bias field (positive bias on the top contact) in Figs. 5 and 6 for the SiO_2 and Al_2O_3 dielectrics, respectively. In Fig. 5 it can be seen that the positive bias will accumulate electrons at the 4H-SiC/SiO_2 interface, where Fowler Nordheim tunneling moves electrons across the relatively large barrier (see Table I). These electrons are then transported through the oxide to the top contact by the Frenkel-Poole mechanism, which lowers the barrier for electron emission from electronic states in the oxide bandgap. Around an electric field of 9 MV/cm the dielectric breakdown occurs and the electrons will create a localized conductive path through the oxide. Figure 6 shows the corresponding IV curve for a typical aluminum oxide MIS structure measured at room temperature. Here, a more rapid increase in the current can be seen starting already from 2.5 MV/cm, probably due to Fowler Nordheim tunneling of electrons through lower barrier between the 4H-SiC/Al_2O_3 interface. This type of behavior has also been documented by others [26]. From about 5 MV/cm, phonon-assisted tunneling is the dominating transport mechanism [23] until the dielectric breakdown. Possibly also Frenkel Poole emission is involved in the breakdown, as seen by the increased current between an electric field of 8 and 9 MV/cm. It appears that the lower offset between Al_2O_3 and 4H-SiC conduction bands, compared to the offset for SiO_2 and 4H-SiC conduction bands, may be a limiting parameter for the utilization of Al_2O_3 directly on the 4H-SiC surface. A thin SiO_2 layer could increase the barrier and improve the breakdown properties, but the purity and homogeniety of this layer must be better than what is typically achieved today.

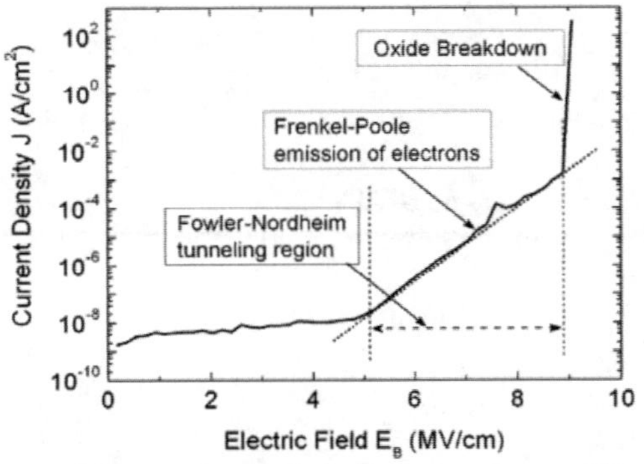

Figure 5. Forward room temperature IV measurements up to the dielectric breakdown of a 4H-SiC MIS structure with SiO_2 deposited by PECVD and post-oxide annealed in N_2O at 1150 ºC. The breakdown is preceded by a wide range of Frenkel-Poole emission of electrons via oxide defects.

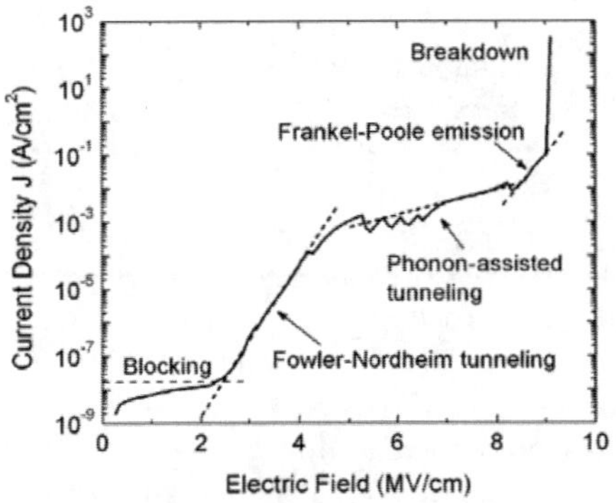

Figure 6. Forward IV measurements at room temperature of a MIS structure consisting of a 4H-SiC epitaxial layer onto which a layer of Al_2O_3 is deposited by ALD.

4. Conclusions

In this paper we have summarized some results of passivation of 4H-SiC surfaces using Al_2O_3 as a alternative dielectric to the prevailing SiO_2. It has been shown that the state of the surface prior to deposition of a dielectric layer is crucial to the final results and basic processing steps, such as cleaning of samples, may considerably influence the interface properties. By introducing an extra cleaning step to remove carbon related contaminants and native oxides (Procedure B), it was possible to obtain higher dielectric constant, lower flatband voltage shift and reduced concentration of interface states. The radiation hardness of Al_2O_3 makes it highly attractive for SiC device applications in radiation rich environments, but the reverse bias leakage currents may still be a problem in high voltage applications.

5. References

[1] Lee H-S, Domeij M, Zetterling C M, Östling M, Allerstam F and Sveinbjörnsson E Ö, 2008 *Appl. Phys. Lett.* **92** 082113
[2] Hallén A, Nawaz M, Zaring C, Usman M, Domeij M, and Östling M, 2010 *IEEE Electr. Device Lett.* **31** 707
[3] Afanasev V V, Bassler M, Pensl G and Schulz M 1997 *Phys. Stat. Sol. (a)* **162**, 321
[4] Rudenko T E, Ólafsson H Ö, Sveinbjörnsson E Ö, Osiyuk I P and I P Tyagulski, 2004 *Microelectronic Engineering* **72**, 213
[5] Virojanadara C and Johansson L I, 2004 *J. Phys.: Condens. Matter* **16**, S3435
[6] Pippel E, Woltersdorf J, Ólafsson H Ö and Sveinbjörnsson E Ö, 2005 *J. Appl. Phys.* **97**, 034302
[7] C. Virojanadara and L.I. Johansson 2005 *Phys. Rev. B* **71**, 195335
[8] Singh R 2006 *Microelectronics and Reliability* **46**, 713
[9] Wolborski W, Bakowski M and Schöner A 2007 *Electronics Letters* **43**, 130

[10] Kosugi R, Ichimura S, Kurokawa A, Koike K, Fukuda K, Suzuki S, Okushi H, Yoshida S and Arai K 2000 *Appl. Surf. Science* **159-160**, 550
[11] Afanas'ev V V, Stesmans A, Bassler M, Pensl G, Schulz M J and Harris C I 1996 *Appl. Phys. Lett.* **68**, 2141
[12] Frank T, Beljakowa S, Pensl G, Kimoto T and Afanas'ev V 2007 *Materials Science Forum* **556-557**, 555
[13] Dhar S, Ryu S-H and Agarwal A 2010 *IEEE Tran. Electr. Dev.* **57**, 1195
[14] Pintille I, Teodorecu C-M, Moscatelli F, Nipoti R, Poggi A, Solmi S, Lövlie L and Svensson B G 2010 *J. Appl. Phys.* **108**, 024503
[15] Wang Y, Khan T and Chow T P 2009 *Materials Science Forum* **600-603**, 751
[16] Seyller T 2004 *J. Phys.: Condens. Matter* **16**, S1755
[17] Palmieri R, Radtke C, Boudinov H and da Silva Jr E F 2009 *Appl. Phys. Lett.* **95**, 113504
[18] Yano H, Katafuchi F, Kimoto T and Matsunami H 1999 *IEEE Transactions on Electron Devices* **49**, 504
[19] Ciobanu F, Pensl G, Afanas'ev V V, Schöner A 2005 *Materials Science Forum* **483-485**, 693
[20] McDonald K, Weller R A, Pantelides S T, Feldman L C, Chung G Y, Tin C C and Williams J R 2003 *J. Appl. Phys.* **93**, 2719
[21] Dimitriev S et al. 2003 **43**, 225 (2003)
[22] Zetterling C M, Östling M, Wongchotigul K, Spencer M G, Tang X, Harris C I, Nordell N and Wong S S 1997 *J. Appl. Phys.* **82**, 2990
[23] Wolborski M, Rosén D, Hallén A and Bakowski M 2006 *Thin Solid Films* **515**, 456
[24] Kolodzey J, Chowdhury E A, Adam T N, Rau I, Olowolafe J O, Suehle J S and Chen Y 2000 *IEEE Transactions on Electron Devices* **47**, 121
[25] Avice M, Grossner U, Pintillie I, Svensson B G, Servidori M, Nipoti R, Nielsen O, Fjellvåg H 2007 *J. Appl. Phys.* **102**, 054513
[26] Wolborski M, Bakowski M, Ortiz A, Pore V, Schöner A, Ritala M, Leskelä M and Hallén A 2006 *Microelectronics Reliability* **46**, 743
[27] Buckley J, De Salvo B, Deleruylle D, Gely M, Nicotra G, Lombardo S, Damlencourt J F, Hollinger Ph, Marti F and Deleonibus S 2005 *Microelectronic Engineering* **80**, 210
[28] Tang X Y, Song Q W, Zhang Y M, Zhang Y M, Jia R X, Lü H L and Wang Y H 2012 *Chin. Phys. B* **21**, 087701
[29] Afanas'ev V V, Stesmans A, Chen F, Cambell S A and Smith R 2003 *Appl. Phys. Lett.* **82**, 922
[30] Wolborski M, Rooth M, Bakowski M and Hallén A 2007 *J. Appl. Phys.* **101**, 124105
[31] Usman M, Hallén A, Pilvi T, Schöner A and Leskelä M 2010 *J. Electrochemical Soc.* **158**, H75-H79
[32] Usman M and Hallén A 2011 *IEEE Electr. Dev. Lett.* **32**, 1653
[33] Usman M, Hallén A, Gulbinas K and VGrivickas V 2012 *Materials Science Forum* 2012
[34] Usman M 2012 PhD Thesis, ISBN 978-91-7501-225-4, Royal Institute of Technology KTH

IOP Conf. Series: Materials Science and Engineering **56** (2014) 012008 doi:10.1088/1757-899X/56/1/012008

Add/drop filters based on SiC technology for optical interconnects

M. Vieira[1,2,3], M. A Vieira[1,2], P. Louro[1,2], A. Fantoni[1,2], V. Silva[1,2]

[1]Electronics Telecommunication and Computer Dept. ISEL, R. Conselheiro Emídio Navarro, 1949-014 Lisboa, Portugal Tel: +351 21 8317290, Fax: +351 21 8317114, mv@isel.ipl.pt ;
[2] CTS-UNINOVA, Quinta da Torre, Monte da Caparica, 2829-516, Caparica, Portugal.
[3] DEE-FCT-UNL, Quinta da Torre, Monte da Caparica, 2829-516, Caparica, Portugal
mv@isel.ipl.pt

Abstract. In this paper we demonstrate an add/drop filter based on SiC technology. Tailoring of the channel bandwidth and wavelength is experimentally demonstrated. The concept is extended to implement a 1 by 4 wavelength division multiplexer with channel separation in the visible range. The device consists of a p-i'(a-SiC:H)-n/p-i(a-Si:H)-n heterostructure. Several monochromatic pulsed lights, separately or in a polychromatic mixture illuminated the device. Independent tuning of each channel is performed by steady state violet bias superimposed either from the front and back sides. Results show that, front background enhances the light-to-dark sensitivity of the long and medium wavelength channels and quench strongly the others. Back violet background has the opposite behaviour. This nonlinearity provides the possibility for selective removal or addition of wavelengths. An optoelectronic model is presented and explains the light filtering properties of the add/drop filter, under different optical bias conditions.

1. Introduction

In the near future, increases in power efficiency per data bit is intended to be achieved by replacing electrical interconnects with their optical counterparts. So, in the visible range, the conception of new devices based on new materials for optically switchable multilayer photonic structures is a demand[1, 2]. A combined RGBV (red, green, blue and violet) wavelength router/switch based on SiC technology can increase, in visible range, the connectivity of an optical system and reduce the number of signal demultiplexers. Tunable backgrounds filters based on a-SiC:H multilayered devices represent a class of filters with characteristics very similar to those of microring resonators [3, 4]. Their filter characteristics make these devices particularly attractive for integrated optics or photonics applications, channel dropping filters, WDM demultiplexers, notch filters and for optical switching or optical wavelength conversion [5]. Filter applications like add/drop multiplexers will be covered in this paper.

2. Device design, characterization and operation

The wavelength add/drop filter is realized by using double pin/pin a-SiC:H photodetector sandwiched between two transparent TCO contacts and two front and back biased optical gating

IOP Conf. Series: Materials Science and Engineering 56 (2014) 012008 doi:10.1088/1757-899X/56/1/012008

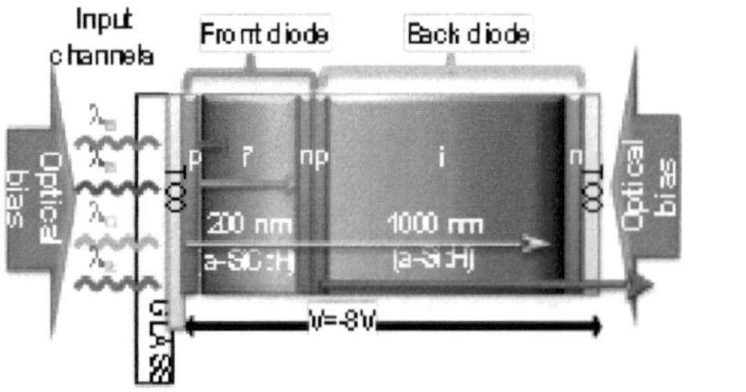

Figure 1 a) Device configuration and operation.

elements as depicted in Figure 1.

The filter consists of a p-i'(a-SiC:H)-n/p-i(a-Si:H)-n heterostructure with low conductivity doped layers. The thicknesses and optical gap of the front i'- (200 nm; 2.1 eV) and back i- (1000 nm; 1.8 eV) layers are optimized for light absorption in the blue and red ranges, respectively [6, 1]. Monochromatic pulsed lights from commercial LED's (input channels) separately (R: 624 nm, 25 μW/cm^2; G: 526 nm, 46 μW/cm^2; B: 470 nm, 60 μW/cm^2; V: 400 nm, 150 μW/cm^2) or in a polychromatic mixture (multiplexed signal) at different bit rates illuminated the device. Independent tuning of each channel is performed by steady state violet optical bias superimposed separately from both sides and the generated photocurrent measured at -8V.The spectral sensitivity, under violet background and without it, was tested through spectral response measurements applied from both sides (Figure 2). In Figure 2a the optical bias was applied from the front side and in Figure 2b from the back side. For comparison the normalized spectral photocurrent for the front, p-i'-n, and the back, p-i-n, photodiodes (dash lines) are superimposed. Those measurements were obtained using two separate similar pin diodes produced in the same conditions as the global structure.

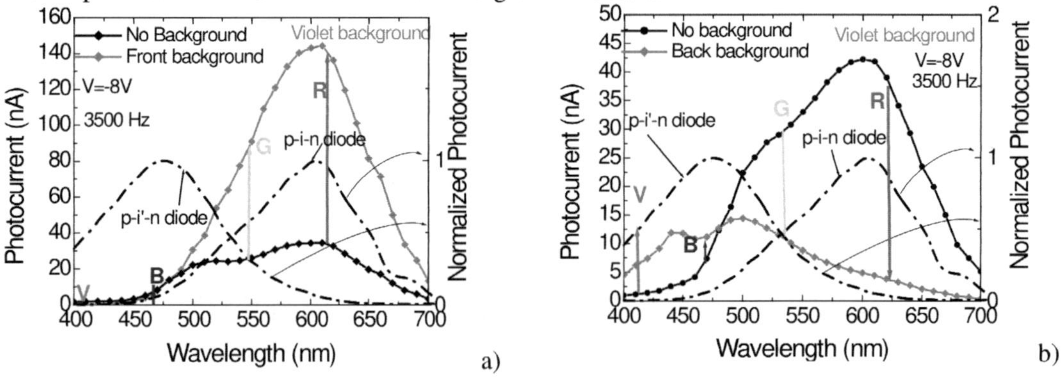

Figure 2 Photocurrent without and with front (a) and back (b) violet background. The normalized photocurrent of the individual photodiodes is superimposed (dash lines).

Data shows that the front and back diodes, separately, presents the typical responses of single p-i-n cells with intrinsic layers based on a-SiC:H or a-Si:H materials, respectively. The front diode cuts the wavelengths higher than 550 nm while the back one rejects the ones lower than 500 nm. The overall device presents an enlarged sensitivity when compared with the individual ones. Under front irradiation the sensitivity is much higher than under back irradiation. Under front irradiation the violet background enhances the spectral sensitivity mainly in the long wavelength range (>550 nm) while back irradiation strongly quenches this and enhances the short wavelength range (see arrows in both figures). Thus, back irradiation, tunes the front diode while front irradiation selects the back one.

3. Optical bias filtering effects

In Figure 3 the spectral gain (α^V), defined as the ratio between the spectral photocurrents under violet illumination and without it, is plotted. The background intensity was changed between 2300μWcm^{-2} and 16μWcm^{-2}.

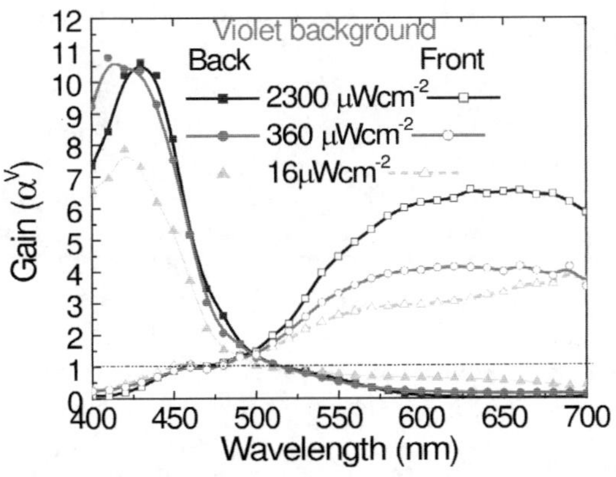

Figure 3 Spectral gain (α^V) under front and back bias.

Results show that under back bias the gain is high at short wavelengths and strongly lowers for wavelengths higher than 500 nm, acting as a short-pass filter. Under violet front light the device works as a long-pass filter for wavelengths higher than 500 nm, blocking the shorter wavelengths. So, by switching the irradiation side the short-, and long- spectral region can be sequentially tuned. The medium region (475 nm-530 nm) can only be tuned by using both active filters. A trade-off between the background side and intensity has to be established. Under front illumination the reddish part of the spectrum is enhanced with the intensity while under back illumination the main enhancement occurs at the violet-blue region.

Four monochromatic pulsed lights separately (red, green, blue and violet input channels, Figure 4) or combined (multiplexed signal; Figures 5 and 6) illuminated the device at 12000 bps. Steady state violet bias was superimposed separately from the front (a) and the back (b) sides and the photocurrent was measured. In Figure 4, the transient signals were normalized to their values without background and, for each individual channel, the mean values of the optical gains added.

Results show that, even under transient conditions and using commercial LED the background side alters, in a different way, the signal magnitude of the input channels. This non-linearity provides the possibility for selective removal or addition of wavelengths.

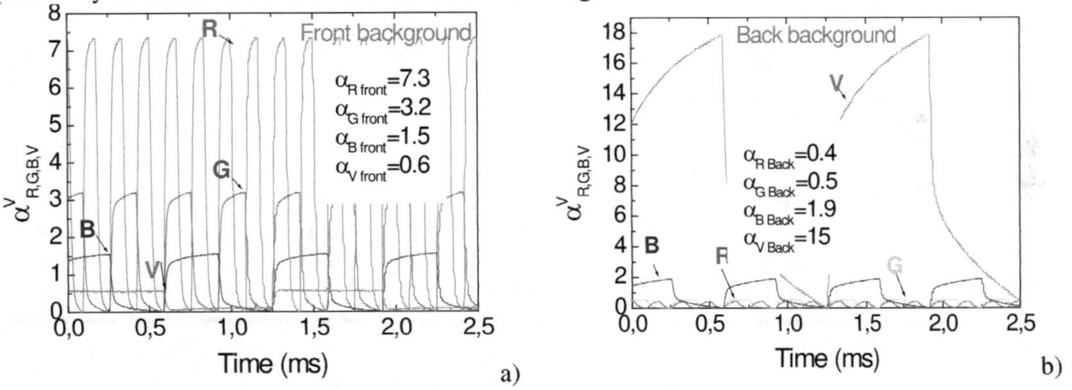

Figure 4 Normalized red, green, blue and violet transient signals at -8V with violet (400 nm) steady state optical bias applied from the front (a) and from the back (b) sides.

As expected from Figure 3, under front irradiation it enhances mainly the spectral sensitivity in the medium-long wavelength ranges ($\alpha^V{}_R$=7.3, $\alpha^V{}_G$=3.2). Violet radiation is absorbed at the top of the front diode (Figure 1), increasing the electric field at the back diode [2] where the red and part of the green incoming photons are absorbed. Under back irradiation the electric field increases mainly near the front p-i´-n interface where the violet and part of the blue incoming channels generate most of the photocarriers ($\alpha^V{}_V$=15, $\alpha^V{}_B$=1.9). So, by switching between front to back irradiation the photonic function is modified from a long- to a short-pass filter allowing, alternately selecting the red or the violet channels. The implementation of a wavelength division multiplexer requires that the channel wavelength may be tailored such that add/drop filters with different spectral locations, may be

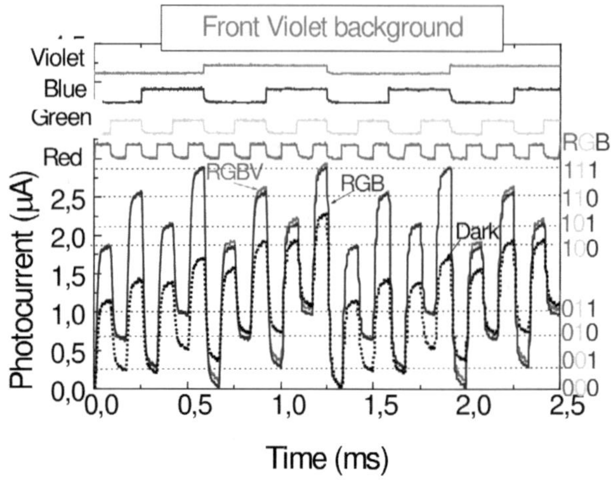

Figure 5 Multiplexed signal under front violet irradiation and without it.

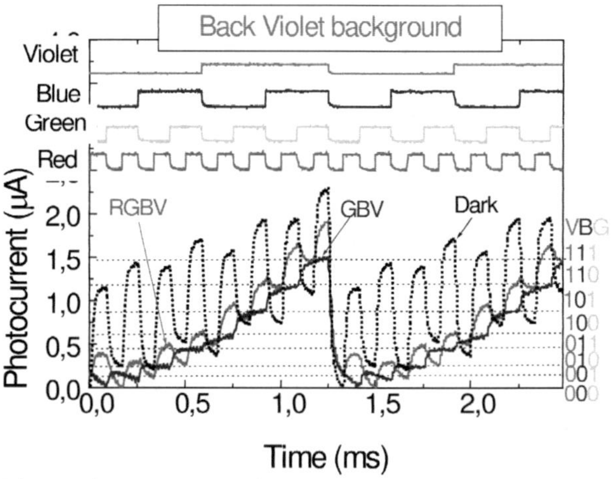

Figure 6 Multiplexed signal under back violet irradiation and without it.

cascaded. Here, tailoring the filter wavelength is achieved by changing the violet background from the front to the back side. The MUX signals due to the combination of the input channels of Figure 4 (R, G, B, V) are displayed without and under front (Figure 5) and back (Figure 6) violet irradiations. On the top the signals used to drive the input channels are displayed showing the presence of all possible sixteen *on/off* states. To show the backgrounds add/drop filter effect, in Figure 5 the RGB and in Figure 6 the VBG combined multiplexed signal are shown. Results show that the device is a two tunable background connected filter and acts as a wavelength selective switch. Without applied background (dark) sixteen output levels are observed each one correspondent to a one of the possible 4-channel *on/off* configurations (16-to-1 multiplexer).

Under applied optical bias, the side of background affects the form and the magnitude of the MUX signal in opposite ways. Under front irradiation the violet channels quenches, and so, the difference between the multiplex signals with and without the violet channel *on* is irrelevant. This means that under front irradiation the transmission of the violet channel changes from high to low and so, the output signal has only eight separate levels (horizontal dot lines) that can be assigned to the 2^3 *on/off* possible states of an 8-to-1 RGB multiplexer. The levels are weighted by the optical gains of each channel (α^V, Figure 4), so, the selection index for this 8-element look-up table (in the right side of Figure 5) is a 3-bit binary [RGB] code. The violet channel was switched to the drop port and the others continue in the transmission line. Under back irradiation the magnitude of the red and green channels are strongly reduced. The 8-to-1 look up table (dot lines) is clear and the 3-assigned bit binary [VBG] code displayed in the right side of Figure 6. So, the quench of the red channel allows it drops to the drop port and the others continue in the transmission line. Similar explanation could be done with the quenching of the green channel resulting in a similar [VBR] code. So, by using the add/drop SiC based filter a usable portion of the optical visible spectrum can be partitioned into wavelength bands to make signal separation possible at the receivers.The advantage of optic background tuning lies in its structural simplicity. The front background selects the red channel and drops the violet while the back background adds the violet and drops the red or the green. It provides a low-cost solution in optical and optoelectronic interconnection technologies according to their suitability of adoption in optical cross connections, switches and routers.

4. Coder/decoder device

The add/drop filter concept is extended to implement a 1 by 4 WDM. Results have shown that under front or back irradiation, each of the four channels, in turn, is enhanced or quenched. The logic MUX function is converted into a logic filter function and each channel can be decoded as a data signal to be routed to a selected destination. Figure 7 displays the normalized MUX signals due to the combination of the input channels of Figure 4. On the top, the signals used to drive the input channels are shown.

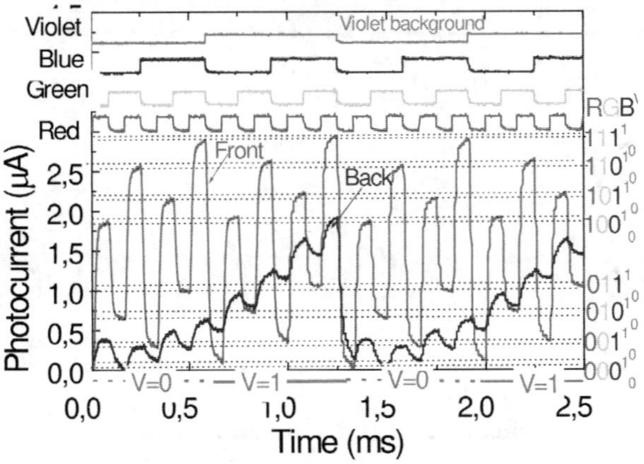

Figure 7 Normalized multiplexed signal under front and back violet irradiation. On the top the signals used to drive the input channels are shown.

Under front irradiation, the 2^4 levels are detected (horizontal dot lines) and grouped into two main classes due to the high amplification of the red channel (Figure 4a). The upper eight (2^3) levels are ascribed to the presence of the red channel (R=1), and the lower eight to its absence (R=0), allowing the red channel decoder (8-to-1 multiplexer; long-pass filter function). Since under front irradiation the green channel is also amplified the four (2^2) highest levels, in both classes, are ascribed to the presence of the green channel (G=1) and the four lower ones to its lack (G=0). The blue channel is slightly amplified, so, in each group of 4 entries, two subclasses (2^1) can be found: the two higher levels correspond to the presence of the blue channel (B=1) and the two lowers to its absence (B=0). Finally, each group of 2 entries have two very near sublevels, the higher where the violet channels is ON (V=1) and the lower where it is missing (V=0). Under back irradiation, the violet channel is strongly enhanced, the blue channel slightly and the green and red reduced (Figure 4b). The encoded multiplexed signal also has sixteen sublevels but ordered in a different way. Two classes are observed, one composed by the eight higher levels, where the violet channel is ON (V=1) and the other including the eight lower levels where it is OFF (V=0) (8-to-1 multiplexer; short-pass filter function) allowing the violet channel decoder (see bottom of Figure 7). Each group the eight levels can still be grouped in two classes of four levels each, with and without the blue channel ON. Taking into account Figure 6, those four near sublevels are ordered. The higher has both red and green channels ON, the lower both OFF and in-between only one of them is missing, the red or the green.

So, the multiplexer select code, under front irradiation, represents an address, *RGBV*, into the ordered inputs while under back irradiation it will be of the form *VBRG*. We may view those 16-element look-up tables, as consisting of two look-ups, one to select the proper group of 8, and pick the red under front irradiation or the violet under back background. Each group of 8 inputs requires 3 bits for picking the proper group of 8 and for specifying an input. By using both look-up tables any multiplex signal can be decoded.

5. Optoelectronic model

Based on the experimental results and device configuration an optoelectronic model, made out of a short- and a long-pass filter was developed [2]. In Figure 8a the *ac* equivalent circuit and the block diagram of the optoelectronic state model are displayed. In Figure 8b the linearized state equations are shown. The use of amplifying elements (Q_1, Q_2), with resistors (R_1, R_2) and capacitors (C_1, C_2) in their feedback loops, synthesize the desired filter characteristics with $\alpha_{R,G,B,V}$ the optical gain.

The input signals, $\lambda_{R,G,B,V}$ model the color channels and *i(t)* the output signal. The amplifying elements, α_1 and α_2 are related with the optical gains and provide gain if needed or attenuate unwanted wavelengths. The control matrix takes into account the enhancement or quenching of the channels

(Figure 4) due to the steady state irradiation. Under front irradiation: $\alpha_2 >> \alpha_1$ and under back irradiation $\alpha_1 >> \alpha_2$. This affects the reverse photo capacitances, ($\alpha_{1,2} / C_{1,2}$) that determine the influence of the system input on the state change (control matrix).

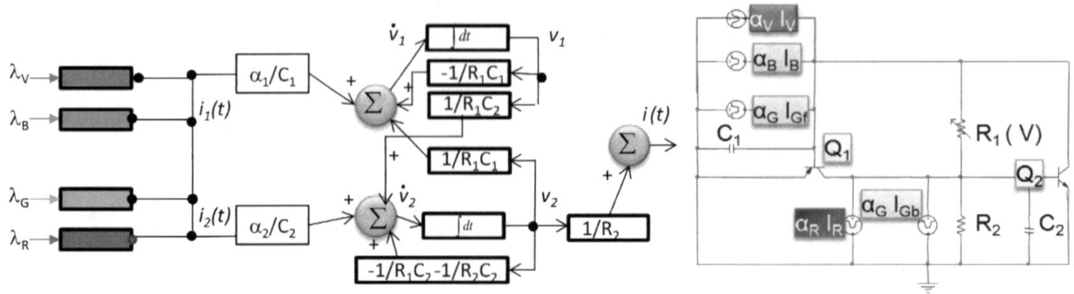

a)

$$\frac{dv_{1,2}}{dt} = \begin{bmatrix}\frac{\alpha_1(\alpha_{V,B,G})}{C_1}\\ \frac{\alpha_2(\alpha_{R,G})}{C_2}\end{bmatrix} i_{1,2}(t) + \begin{bmatrix} -\frac{1}{R_1C_1} & \frac{1}{R_1C_1}\\ \frac{1}{R_1C_2} & -\frac{1}{R_1C_2}-\frac{1}{R_2C_2}\end{bmatrix} v_{1,2}(t) \;\; ; i_{1,2}(t) = \begin{bmatrix}0 & \frac{1}{R_2}\end{bmatrix} v_{1,2}(t)$$

b)

Figure 8. a) Block diagram and *ac* equivalent circuit. b) Linearized state equations.

The opto-electrical model with light biasing control has proven to be a good tool to design SiC multilayer add/drop filters. Furthermore, this model allows for extracting experimental parameters by fitting the model to the measured data.

6. Conclusions

An add/drop filter based on a-SiC:H technology was analyzed. Spectral response measurements were presented.

Results have been shown that tailoring the filter wavelength was achieved by applying a 400 nm steady state optical bias either from the front or from the back sides. The add/drop filter concept was extended to implement a 1 by 4 WDM. Under front or back irradiation, each of the four channels, in turn, is enhanced or quenched. The logic MUX function was converted into a logic filter function and each channel decoded as a data signal to be routed to a selected destination.

An optoelectronic model was present and explains the add/drop filtering properties. The opto-electrical model has proven to be a good tool to design SiC multilayer add/drop filters.

References

[1] M. Vieira, P. Louro, M. Fernandes, M. A. Vieira, A. Fantoni and J. Costa 2011 *Advances in Photodiodes, InTech*, **Chap.19**, pp:403-425-

[2] M. A. Vieira, M. Vieira, J. Costa, P. Louro, M. Fernandes, A. Fantoni, 2010 *Sensors & Transducers Journal*, **9**, Special Issue, pp.96-120.

[3] P. Dong, R. Shafiiha, S. Liao, H. Liang, N.-N. Feng, D. Feng, G. Li, X. Zheng, A. V. Krishnamoorthy, and M. Asghari 2010 *Opt. Express* **18 (11),** pp.10941–10946.

[4] L. Zhang, R. Q. Ji, L. X. Jia, L. Yang, P. Zhou, Y. H. Tian, P. Chen, Y. Y. Lu, Z. Y. Jiang, Y. L. Liu, Q. Fang, and M. B. Yu 2010 *Opt. Lett.* **35(10)**, 1620–1622.

[5] C. Petit, M. Blaser, 2006, *Workshop on Optical Components for Broadband Communication* , ed. by Pierre-Yves Fonjallaz, Thomas P. Pearsall, Proc. of SPIE **6350**, 63500I.

[6] M.A. Vieira, P. Louro, M. Vieira, A. Fantoni, A. Steiger-Garção. 2012 IEEE *sensor jornal,* **12**, NO. 6, pp. 1755-1762.

IOP Conf. Series: Materials Science and Engineering **56** (2014) 012009 doi:10.1088/1757-899X/56/1/012009

Structural and luminescence properties of undoped and Eu-doped SiOC thin films

G Bellocchi [1,2], G Franzò[1], S Boninelli[1], M Miritello[1], T Cesca[3], F Iacona[1] and F Priolo[1,2]

[1] MATIS IMM CNR, via Santa Sofia 64, 95121 Catania, Italy.
[2] Dipartimento di Fisica e Astronomia, Università di Catania, via Santa Sofia 64, 95121 Catania, Italy.
[3] Dipartimento di Fisica e Astronomia, Università di Padova, Via F. Marzolo 8, 35131 Padova, Italy

E-mail: gabriele.bellocchi@ct.infn.it

Abstract. We report on the investigation of the structural, chemical and optical properties of undoped and Eu-doped SiOC thin films synthetized by RF magnetron sputtering. Undoped SiOC exhibits an intense room temperature luminescence at ~ 500 nm, and an important contribution to this signal is due to the presence of Si-C bonds. Moreover, when inserted in a SiOC matrix, Eu ions have a higher solid solubility with respect to pure SiO_2; as a consequence, we observe a reduction of Eu clustering phenomena. Furthermore the reducing properties of SiOC, related to the presence of C, allow the prevalence of the Eu^{2+} emission over the Eu^{3+} one. Through the control of the annealing conditions it is possible to obtain an intense light emission at 440 nm. These properties propose SiOC as a novel and efficient Si-based host matrix for Eu and open the way to promising perspectives of Eu-based materials for photonic applications and LED fabrication.

1. Introduction

The realization of an efficient Si-based light source represents one of the most relevant issues for the scientific community currently working in the field of photonics. The main characteristics required for this source are high efficiency at room temperature, tunability in a wide wavelength range, easy integration in highly compact Si microphotonic devices. In this framework, a leading role is played by rare earth (RE) doped materials. RE doping of insulating or semiconducting host matrices has been recognized as a powerful method to provide intense luminescence signals spanning in a wide wavelength range, opening the way to several new applications for Si-based materials in the field of photonics [1]. Among the RE ions, a considerable attention has been devoted to Eu due to the coexistence of both divalent and trivalent oxidation states. Indeed Eu^{3+} and Eu^{2+} ions can act as efficient emitting centers in the visible region and their bright emission is widely employed in phosphors used in plasma display panels and in solid-state lighting [2]. Divalent Eu ions exhibit a very strong emission originating from dipole-allowed $4f^65d \rightarrow 4f^7$ transitions, favorite with respect to that of Eu^{3+} ions, whose intra 4f-shell transition, being electric dipole-forbidden, leads to a less intense luminescence [3]. On the other hand, applications of Eu in Si photonics require the availability of Eu-containing thin films compatible with CMOS technology. Several papers have studied the optical properties of Eu-doped SiO_2 films [4-7], but the low Eu solid solubility in silica (cluster formation occurs for Eu concentrations larger than 10^{18} cm^{-3} [8-10]) seems to seriously limit its perspectives in photonics.

A different approach to obtain efficient light emission from a Si-based material is represented by Si oxycarbide (SiOC), which has been extensively studied for its intense photoluminescence (PL) emission in the visible range [11-17]. Furthermore, it has been also evidenced that SiOC can be an interesting host material for RE ions, including Er [18] and Eu [10,19].

In this paper we will study the optical properties of SiOC and Eu-doped SiOC thin films synthesized by RF magnetron sputtering. It will be demonstrated that the temperature and the environment of the thermal annealing process following the deposition can improve the PL emission in the visible region from both materials. The mechanisms of the SiOC-Eu interaction, leading to a very bright visible emission assigned to Eu^{2+} ions, will be also elucidated. The relevance of the above data for the realization of an efficient Si-based light source, which can be of great interest for applications in photonics as well as in solid-state lighting, will be addressed.

2. Experimental

SiOC and Eu-doped SiOC thin films, ~190 nm thick, have been synthesized by using an UHV RF magnetron sputtering system. The chamber base pressure was ~ 1×10^{-9} mbar. The deposition has been obtained by co-sputtering of 4" diameter, water-cooled SiO_2 and SiC targets (in the case of SiOC samples), and SiO_2, SiC and Eu_2O_3 targets (in the case of Eu-doped SiOC samples). C concentration has been fixed at 5 at.% in both doped and undoped SiOC films, while Eu concentration was 1.5×10^{20} cm^{-3}. The films were deposited on Si substrates kept at 400 °C in a 5×10^{-3} mbar Ar atmosphere. After deposition the films were thermally treated at temperatures ranging from 750 to 1000 °C for 1 h in N_2 atmosphere.

Transmission electron microscopy (TEM) analysis in bright field (BF) mode, performed by a 200 kV Jeol 2010 microscope, was used to investigate the film structure. The atomic composition of the films was studied by Rutherford backscattering spectrometry (RBS), using a 2 MeV He^+ beam in random configuration, with a detector placed at 165° with respect to the incident beam.

PL measurements were performed by pumping with the 325 nm line of a He-Cd laser. The pump power was about 3 mW and the laser beam was chopped through an acousto-optic modulator at a frequency of 55 Hz. The PL signal was analyzed by a single grating monochromator and detected with a photomultiplier. Spectra were recorded with a lock-in amplifier using the acousto-optic modulator frequency as a reference. Photoluminescence excitation (PLE) measurements in the 250-475 nm range were performed by using a FluoroMax spectrofluorometer by Horiba. All the spectra have been measured at room temperature and corrected for the system spectral response.

3. Results and discussion

The room temperature PL spectra of as-deposited and thermally annealed SiOC films are reported in figure 1.

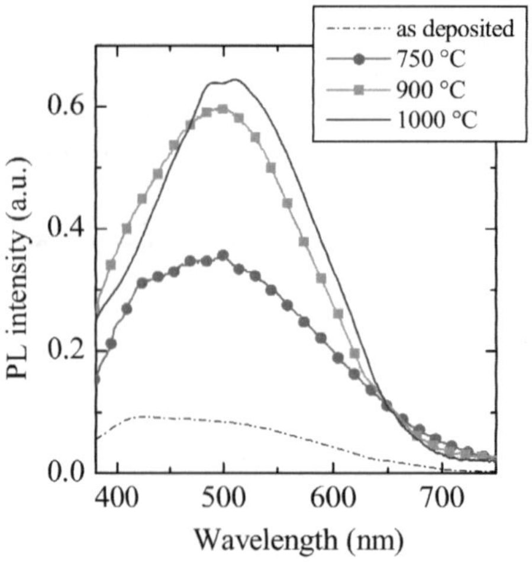

Figure 1. Room temperature PL spectra of SiOC films as deposited and annealed in N_2 ambient in the temperature range 750 – 1000 °C.

These measurements show that SiOC films are efficient light emitters at room temperature in the visible region. It is evident that the temperature of the thermal treatment strongly influence the PL properties. Spectra of samples annealed at high temperatures are generally red-shifted and sharper with respect to those of the as-deposited or low temperatures annealed samples. More in detail, the spectrum of the as-deposited sample appears quite broad, suggesting the presence of multiple luminescent centers. After the annealing process, by increasing the temperature, a component at 510 nm becomes predominant, although it is still visible a shoulder at around 400 nm. Since in annealed samples a partial ordering of the film structure occurs (leading to the formation of optically active Si-C bonds and to the quenching of defects), the evidence that the system becomes more optically efficient at higher temperatures suggests that the PL at 510 nm can be assigned to Si-C bonds, while the PL at 400 nm is due to defect centers.

As previously reported [10,19-21], SiOC can be also a very suitable host matrix for optically active rare earth ions. In particular, Eu precipitation, which strongly limits the optical activity of Eu ions embedded in the more conventional SiO_2 matrix, can be reduced when incorporated in SiOC.
Figure 2 reports the BF XTEM images relative to Eu-doped SiOC films annealed at 900 °C (a) and 1000 °C (b) in N_2 ambient. The Eu concentration is 1.5×10^{20} cm^{-3} for both samples. From the analysis of the figure it is evident that when SiC is added to the SiO_2 matrix, in spite of the fact that the Eu concentration is about 2 orders of magnitude higher than the threshold for Eu precipitation in pure SiO_2, Eu precipitates are detected only at the sample surface.

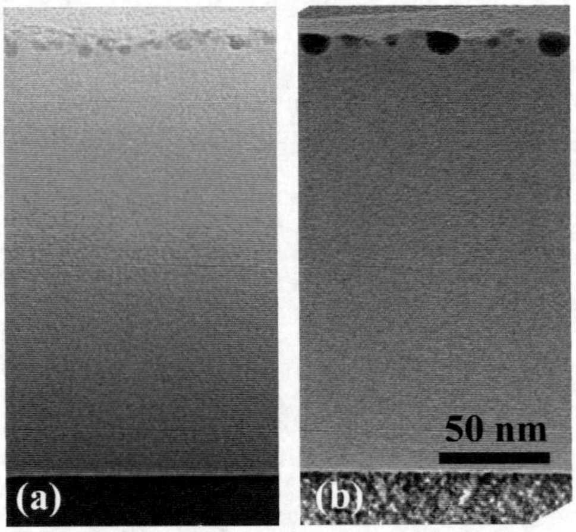

Figure 2. BF XTEM images relative to Eu-doped SiOC layers annealed at 900 °C (a) and 1000 °C (b) in N_2 ambient. The Eu concentration is 1.5×10^{20} cm^{-3} in both samples.

In order to understand why Eu precipitates are formed at the sample surface, we have performed RBS measurements to follow the evolution of the Eu profile with the annealing temperature. In figure 3, the comparison between the Eu concentration profiles relative to an as-deposited Eu-doped SiOC film (black line) and those corresponding to thermally annealed samples (900 °C, red line); 1000 °C, blue line) are shown. The data show an uniform Eu concentration in as-deposited film, while after annealing Eu surface segregation occurs and becomes more evident with increasing temperature. The surface Eu concentration peaks account for about 10% (at 900 °C) and 65% (at 1000 °C) of the total Eu concentration. In spite of the segregation process, a high Eu content (~ $1.5\times10^{20}/cm^3$ at 900 °C and $5.5\times10^{19}/cm^3$ at 1000 °C), well above Eu solid solubility in pure SiO_2, remains dissolved in the SiOC matrix, from 50 nm below the surface, without precipitating.
Furthermore, SiOC is also able to induce the $Eu^{3+}\rightarrow Eu^{2+}$ reduction, providing a simple method to obtain high concentrations of the most optically efficient Eu species (Eu^{2+}). Figure 4 reports the room temperature PL spectra of undoped and Eu-doped SiOC samples, treated at 1000 °C in N_2. The Eu-

doped SiOC (blue line) exhibits a very intense PL band, centered at about 440 nm, due to the $4f^65d \rightarrow 4f^7$ transitions of Eu^{2+} ions, massively formed owing to a reduction reaction involving the C atoms of the SiOC matrix [12]. No sharp lines associated to Eu^{3+} PL emission are detected. The PL signal at 440 nm is 10 times higher than that measured in the undoped SiOC film (red line).

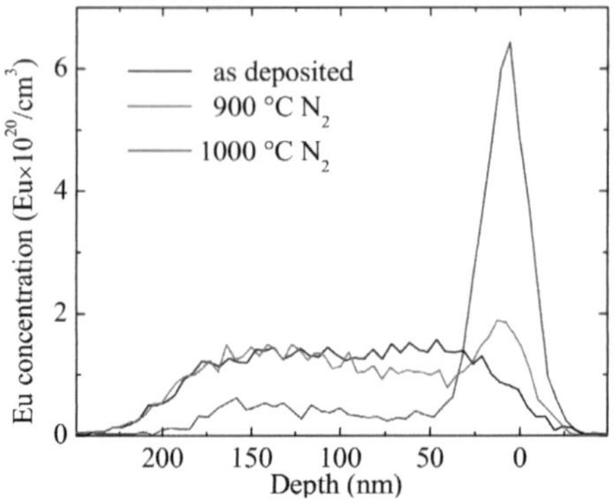

Figure 3. Eu concentration profiles, measured by RBS, relative to Eu-doped SiOC films as-deposited (black) and annealed in N_2 ambient at 900 (red) and 1000 °C (blue).

The integrated PL intensity measured in both undoped and Eu-doped SiOC as a function of the annealing temperature (not shown) shows a similar trend, probably due to the fact that Eu^{2+} ions are formed owing to a redox reaction involving Eu^{3+} ions and Si-C bonds (which give the main contribution to the PL signal of undoped SiOC films annealed at high temperature in N_2). Indeed, by increasing the annealing temperature, the concentration of Si-C bonds increases and, as a consequence, the number of emitting Eu^{2+} ions, formed through the redox reaction, becomes higher.

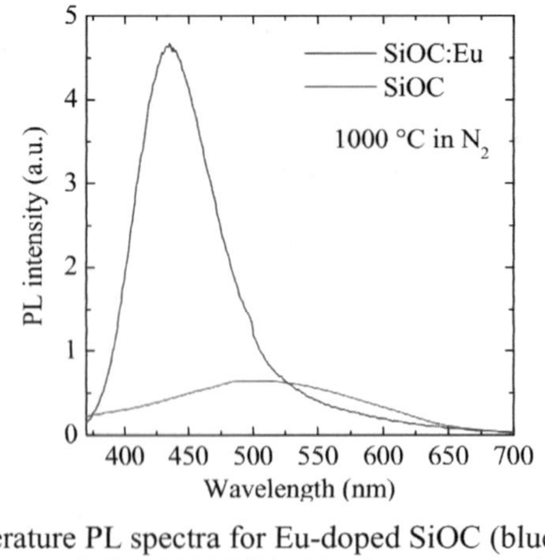

Figure 4. Room temperature PL spectra for Eu-doped SiOC (blue line), and undoped SiOC (red line) films, annealed at 1000 °C in N_2. Eu concentration of the doped film is 1.5×10^{20} cm^{-3}.

Figure 5 reports the room temperature PLE spectra of undoped and Eu-doped SiOC with an Eu concentration of 1.5×10^{20} cm^{-3}, obtained by integrating the PL spectra measured by exciting in the 260-370 nm wavelength range. PLE curve of Eu-doped SiOC (blue line and triangles) exhibits a maximum at 250 nm followed by a monotonic decrease for longer excitation wavelengths. PLE spectrum of undoped SiOC (red line and circles) shows a maximum at 260 nm and monotonically decreases by increasing the excitation wavelength with a trend similar to that of Eu-doped SiOC. From the comparison with the PLE spectrum of an Eu^{2+} silicate (green line and squares) [22], reported in the same figure, it is evident that the PL efficiency of Eu-doped SiOC is strongly enhanced for wavelengths ≤ 300 nm; indeed the Eu^{2+} silicate exhibits a maximum at about 350 nm and it is remarkably less efficient for shorter excitation wavelengths, in agreement with literature data concerning very different Eu^{2+}-containing systems [23-25].

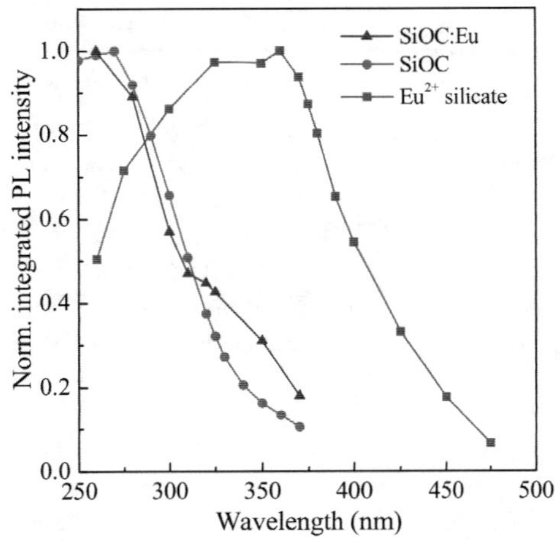

Figure 5. Comparison of the room temperature PLE spectra of undoped and Eu-doped SiOC films and of an Eu^{2+} silicate. The spectra are normalized to their maximum value.

It is possible to conclude that the SiOC matrix has an active role in the optical properties of Eu-doped SiOC. We can reasonably exclude that this role consists in the presence of a SiOC emission, since the spectral shape shown in figure 4 for an excitation wavelength of 325 nm does not appreciably change by changing the excitation wavelength, even by using the wavelengths which maximize a possible SiOC emission. We could therefore suppose the existence of an energy transfer mechanism, in which the very efficient absorption properties of SiOC in the short wavelength region are exploited to transfer the excitation energy to the Eu^{2+} ions, by enhancing their PL efficiency under excitation at short wavelength.

4. Conclusions

The intense visible emission of SiOC layers has been studied and its dependence on the parameters of the thermal annealing process elucidated. In particular, the presence of different emitting centers, and in particular one related to the presence of Si-C bonds, have been identified.

Although the emission of SiOC is bright enough to be considered potentially interesting for applications in photonics (as the active medium in Si-based light emitting devices) or in lighting technology, this material has been found to be even more promising as a host matrix for optically active Eu ions. Indeed a SiOC matrix is able to efficiently promote the $Eu^{3+} \rightarrow Eu^{2+}$ reduction. Furthermore, TEM and RBS data have demonstrated that Eu ions in SiOC are characterized by an enhanced solubility; this peculiarity leads to a strongly reduced Eu precipitation, and, as a consequence, to a very intense and stable light emission at about 440 nm from Eu^{2+} ions, 10 times higher with respect to undoped SiOC. We have also found evidences of the occurrence of an energy transfer mechanism between the matrix and the Eu^{2+} ions which, by increasing the efficiency of photon absorption for excitation wavelengths shorter than 300 nm, further contributes to increase the optical efficiency of Eu-doped SiOC layers. These properties open the way to new promising perspectives for the application of Eu-doped materials in photonics and LED fabrication.

References

[1] Franzò G, Vinciguerra V and Priolo F 1999 *Appl. Phys. A: Mater. Sci. Process.* **69** 3-12

[2] Ye X, Zhuang W, Hu Y, He T, Huang X, Liao C, Zhong S, Xu Z, Nie H and Deng G 2009 *J. Appl. Phys.* **105** 064302

[3] Blasse G and Grabmaier B C 1994 *Luminescent Materials* (Berlin: Springer Verlag)

[4] Li D, Zhang X, Jin L, and Yang D 2010 *Opt. Express* **18** 27191-27196

[5] Rebohle L, Lehmann J, Prucnal S, Kanjilal A, Nazarov A, Tyagulskii I, Skorupa W, and Helm M 2008 *Appl. Phys. Lett.* **93** 071908

[6] Rebohle L, Lehmann J, Prucnal S, Nazarov A, Tyagulskii I, Tyagulskii S, Kanjilal A, Voelskow M, Grambole D, Skorupa W and Helm M 2009 *J. Appl. Phys.* **106** 123103

[7] Prucnal S, Sun J M, Skorupa W and Helm M 2007 *Appl. Phys. Lett.* **90** 181121

[8] Afify N D and Mountjoy G 2009 *Phys. Rev. B* **79** 024202

[9] Laegsgaard J 2002 *Phys. Rev. B* **65** 174114

[10] Boninelli S, Bellocchi G, Franzò G, Miritello M and Iacona F 2013 *J. Appl. Phys.* **113** 143503

[11] Seo S-Y, Cho K-S and Shin J H 2004 *Appl. Phys. Lett.* **84** 717-719

[12] Gallis S, Nikas V, Suhag H, Huang M and Kaloyeros A E 2010 *Appl. Phys. Lett.* **97** 081905

[13] Guo Y P, Zheng J C, Wee A T S, Huan C H A, Li K, Pan J S, Feng Z C and Chua S 2001 *J Chem. Phys. Lett.* **339** 319-322

[14] Perez-Rodriguez A, Gonzalez-Varona O, Garrido B, Pellegrino P, Morante J R, Bonafos C, Carrada M and Claverie A 2003 *J. Appl. Phys.* **94** 254-262

[15] Ishikawa Y, Vasin A V, Salonen J, Muto S, Lysenko V S, Nazarov A N, Shibata N and Lehto V-P 2008 *J. Appl. Phys.* **104** 083522

[16] Ding Y and Shirai H 2009 *J. Appl. Phys.* **105** 043515

[17] Vasin A V, Ishikawa Y, Shibata N, Salonen J, and Lehto V-P 2007 *Jpn. J. Appl. Phys.* **46** L465-L467

[18] Gallis S, Huang M and Kaloyeros A E 2007 *Appl. Phys. Lett.* **90** 161914

[19] Zhang Y, Quaranta A and Soraru G D 2004 *Opt. Mater.* **24** 601-605

[20] Gallis S, Huang M, Efstathiadis H, Eisenbraun E, Kaloyeros A E, Nyein E E and Hommerich U 2005 *Appl. Phys. Lett.* **87** 091901

[21] Seo S-Y and Shin J H 2004 *Appl. Phys. Lett.* **84** 4379

[22] Bellocchi G, Franzò G, Iacona F, Boninelli S, Miritello M, Cesca T and Priolo F 2012 *Opt. Express* **20** 5501-5507

[23] Zhang Q, Liu X, Qiao Y, Qian B, Dong G, Ruan J, Zhou Q, Qiu J and Chen D 2010 *Opt. Mater.* **32** 427-431

[24] Kishimoto Y, Zhang X, Hayakawa T and Nogami M 2009 *J. Lumin.* **129** 1055-1059

[25] Qiao Y, Chen D, Ren J, Wu B, Qiu J, and Akai T 2008 J. Appl. Phys. **103** 023108

AUTHOR INDEX

Institute of Physics
Dirac House, Temple Back
Bristol BS1 6BE UK